Transportation of Laboratory Animals

PROCEEDINGS OF A WORKSHOP

Patricia McAdams, Steven Olson, Lida Anestidou, and Jenna Ogilvie, *Rapporteurs*

Roundtable on Science and Welfare in Laboratory Animal Use

Institute for Laboratory Animal Research

Division on Earth and Life Studies

The National Academies of
SCIENCES · ENGINEERING · MEDICINE

THE NATIONAL ACADEMIES PRESS
Washington, DC
www.nap.edu

THE NATIONAL ACADEMIES PRESS **500 Fifth Street, NW** **Washington, DC 20001**

This project was supported by the American College of Laboratory Animal Medicine; American Veterinary Medical Association; Bayer; Charles River Laboratories; Covance Laboratories, Inc.; GlaxoSmithKline; Johnson & Johnson; Massachusetts General Hospital; Massachusetts Institute of Technology; Merck; Novartis Corporation; University of California, Davis; University of Illinois at Urbana-Champaign; University of Michigan; University of Washington; and the National Academy of Sciences W.K. Kellogg Foundation Fund. Any opinions, findings, conclusions, or recommendations expressed in this publication do not necessarily reflect the views of any organization or agency that provided support for the project.

International Standard Book Number-13: 978-0-309-37333-3
International Standard Book Number-10: 0-309-37333-6
Digital Object Identifier: https://doi.org/10.17226/21734

Additional copies of this publication are available for sale from the National Academies Press, 500 Fifth Street, NW, Keck 360, Washington, DC 20001; (800) 624-6242 or (202) 334-3313; http://www.nap.edu.

Printed in the United States of America

Suggested citation: National Academies of Sciences, Engineering, and Medicine. 2017. *Transportation of Laboratory Animals: Proceedings of a Workshop*. Washington, DC: The National Academies Press. doi: https://doi.org/10.17226.21734.

The National Academies of

SCIENCES • ENGINEERING • MEDICINE

The **National Academy of Sciences** was established in 1863 by an Act of Congress, signed by President Lincoln, as a private, nongovernmental institution to advise the nation on issues related to science and technology. Members are elected by their peers for outstanding contributions to research. Dr. Marcia McNutt is president.

The **National Academy of Engineering** was established in 1964 under the charter of the National Academy of Sciences to bring the practices of engineering to advising the nation. Members are elected by their peers for extraordinary contributions to engineering. Dr. C. D. Mote, Jr., is president.

The **National Academy of Medicine** (formerly the Institute of Medicine) was established in 1970 under the charter of the National Academy of Sciences to advise the nation on medical and health issues. Members are elected by their peers for distinguished contributions to medicine and health. Dr. Victor J. Dzau is president.

The three Academies work together as the **National Academies of Sciences, Engineering, and Medicine** to provide independent, objective analysis and advice to the nation and conduct other activities to solve complex problems and inform public policy decisions. The National Academies also encourage education and research, recognize outstanding contributions to knowledge, and increase public understanding in matters of science, engineering, and medicine.

Learn more about the National Academies of Sciences, Engineering, and Medicine at **www.nationalacademies.org**.

The National Academies of

SCIENCES • ENGINEERING • MEDICINE

Consensus Study Reports published by the National Academies of Sciences, Engineering, and Medicine document the evidence-based consensus on the study's statement of task by an authoring committee of experts. Reports typically include findings, conclusions, and recommendations based on information gathered by the committee and the committee's deliberations. Each report has been subjected to a rigorous and independent peer-review process and it represents the position of the National Academies on the statement of task.

Proceedings published by the National Academies of Sciences, Engineering, and Medicine chronicle the presentations and discussions at a workshop, symposium, or other event convened by the National Academies. The statements and opinions contained in proceedings are those of the participants and are not endorsed by other participants, the planning committee, or the National Academies.

For information about other products and activities of the National Academies, please visit www.nationalacademies.org/about/whatwedo.

PLANNING COMMITTEE FOR TRANSPORTATION OF LABORATORY ANIMALS[1]

Carol Clarke (*Co-Chair*), Research Specialist Staff Officer, U.S. Department of Agriculture, Animal and Plant Health Inspection Service, Animal Care

William J. White (*Co-Chair*), Corporate Vice President, Charles River Laboratories

Robert C. Dysko, Clinical Professor, Unit for Laboratory Animal Medicine, University of Michigan Medical School

Judith B. Franco, Associate Director of Comparative Medicine, Pfizer Inc.

Dianne Garnes, Global Animal Welfare Officer, Novartis Corporation

Bruce W. Kennedy, Compliance Associate, California State Polytechnic University

David M. Kurtz, Veterinary Staff Scientist, National Institute of Environmental Health Sciences

C. Ford Morishita, Retired Biology Teacher, Founding Member of the National Academies Teacher Advisory Council

Susan Brust Silk, Director, Division of Policy and Education, Office of Laboratory Animal Welfare, National Institutes of Health

---

[1] Affiliations of planning committee members were current as of September 2014.

ROUNDTABLE ON SCIENCE AND WELFARE IN LABORATORY ANIMAL USE[1]

Co-Chairs

Lynn C. Anderson (*Co-Chair*), Vice President, Global Animal Welfare and Comparative Medicine, Covance Laboratories, Inc.

Steven Niemi (*Co-Chair*), Liaison to the American College of Laboratory Animal Medicine; and Director, Office of Animal Resources, Harvard University Faculty of Arts and Sciences

Members

Paul A. Locke, Liaison to Institute for Laboratory Animal Research (ILAR) Council; and Associate Professor, Department of Environmental Health Sciences, Johns Hopkins Bloomberg School of Public Health

David Anderson, Executive Director, Health Sciences Administration, University of Washington

Bonnie V. Beaver, Professor, Department of Small Animal Clinical Sciences, College of Veterinary Medicine, Texas A&M University

Pamela Chamberlain, Veterinary Medical Officer and Institutional Official for the White Oak Animal Program, Office of Counterterrorism and Emerging Threats, U.S. Food and Drug Administration

Carol Clarke, Research Specialist Staff Officer, Animal and Plant Health Inspection Service, Animal Care, U.S. Department of Agriculture

Robert C. Dysko, Clinical Professor, Unit for Laboratory Animal Medicine, University of Michigan Medical School

James Fox, Professor and Director, Division of Comparative Medicine, Massachusetts Institute of Technology

Dianne Garnes, Global Animal Welfare Officer, Pharmaceutical Division, Novartis Corporation

Gail C. Golab, Director, Animal Welfare Division, American Veterinary Medical Association

Donna Matthews Jarrell, Attending Veterinarian and Director, Center for Comparative Medicine, Massachusetts General Hospital

[1] Affiliations of roundtable members were current as of September 2014.

Bruce W. Kennedy, Compliance Associate, California State Polytechnic University
David M. Kurtz, Veterinary Staff Scientist, National Institute of Environmental Health Sciences
Margaret S. Landi, Chief of Animal Welfare, Ethics, and Strategy, GlaxoSmithKline
K. C. Kent Lloyd, Professor and Head, Mouse Biology Program, University of California, Davis
Lawrence Schook, Professor, Office of the Vice President for Research, University of Illinois at Urbana-Champaign
Susan Brust Silk, Director, Division of Policy and Education, Office of Laboratory Animal Welfare, National Institutes of Health
William J. White, Corporate Vice President, Veterinary and Professional Services, Charles River Laboratories
John C. Wingfield, Assistant Director, Biological Sciences, National Science Foundation
Robert H. Wurtz, National Institutes of Health Distinguished Investigator, Laboratory of Sensorimotor Research, National Eye Institute, National Institutes of Health

Staff

Lida Anestidou, Director, Roundtable on Science and Welfare in Laboratory Animal Use
Angela Kolesnikova, Administrative Assistant, Institute for Laboratory Animal Research
Jenna Ogilvie, Senior Program Assistant, Board on Life Sciences

INSTITUTE FOR LABORATORY ANIMAL RESEARCH[1]

---

[1] Affiliations of Institute for Laboratory Animal Research (ILAR) Council members were current as of September 2014.

Staff
Frances E. Sharples, Acting Director
Lida Anestidou, Senior Program Officer
Angela Kolesnikova, Administrative Assistant

Reviewers

This Proceedings of a Workshop was reviewed in draft form by individuals chosen for their diverse perspectives and technical expertise. The purpose of this independent review is to provide candid and critical comments that will assist the National Academies of Sciences, Engineering, and Medicine in making each published proceedings as sound as possible and to ensure that it meets the institutional standards for quality, objectivity, evidence, and responsiveness to the charge. The review comments and draft manuscript remain confidential to protect the integrity of the process.

We thank the following individuals for their review of this proceedings:

Anne Deschamps, Federation of American Societies of Experimental Biology
Erik D. Dohm, University of Alabama at Birmingham
Mary A. Robinson, University of Texas Health Science Center at Houston
Joe H. Simmons, University of Texas MD Anderson Cancer Center

Although the reviewers listed above provided many constructive comments and suggestions, they were not asked to endorse the content of the proceedings nor did they see the final draft before its release. The review of this proceedings was overseen by **Allan Basbaum**, University of California, San Francisco. He was responsible for making certain that an independent examination of this proceedings was carried out in accordance with standards of the National Academies and that all review comments were carefully considered. Responsibility for the final content rests entirely with the rapporteurs and the National Academies.

Contents

Boxes

Acronyms and Abbreviations

AALAS	American Association for Laboratory Animal Science
APHIS	Animal and Plant Health Inspection Service
ATA	Animal Transport Association
CBP	U.S. Customs and Border Patrol
CDC	U.S. Centers for Disease Control and Prevention
CITES	Convention on International Trade in Endangered Species of Wild Fauna and Flora
DHS	Department of Homeland Security
DOT	Department of Transportation
EARA	European Animal Research Association
e-CFR	Electronic Code of Federal Regulations
EU	European Union
FDA	U.S. Food and Drug Administration
FWS	U.S. Fish & Wildlife Service
HACCP	Hazard Analysis and Critical Control Point system
IACUC	Institutional Animal Care and Use Committee
IATA	International Air Transport Association
ILAR	Institute for Laboratory Animal Research
LAPB	Live Animals and Perishables Board
LAR	Live Animals Regulations
NC3Rs	National Centre for Replacement, Refinement, and Reduction of Animals in Research
NGSS	Next Generation Science Standards
NHP	nonhuman primate
NIEHS	National Institute of Environmental Health Sciences

NIH	National Institutes of Health
OIE	World Organisation for Animal Health
SPF	specific pathogen free
USDA	U.S. Department of Agriculture
WHO	World Health Organization

1

Introduction and Overview

"Transporting animals safely and humanely is the way to really address the Three Rs."
—William White, Charles River Laboratories

The obligation to treat animals used in research ethically and humanely extends beyond their lives in the laboratory to include their transportation from place to place. Yet transporting animals is a highly regulated and complex process that raises many difficult issues.

To examine these issues, the Roundtable on Science and Welfare in Laboratory Animal Use held a workshop on September 3–4, 2014, in Washington, DC, titled Transportation of Laboratory Animals. More than 200 people participated in the workshop in person and online, including representatives of academic research institutions, pharmaceutical and consumer product companies, government agencies, research advocacy groups, professional associations, and the public.

As co-chair of the roundtable, Lynn Anderson, vice president of animal welfare and comparative medicine at Covance Laboratories, Inc.,[1] said, the ultimate goal of both the workshop and the roundtable is to "foster communication and do problem solving on important issues facing the laboratory animal community." Thus, the workshop was designed to draw attention to the essential, thoughtful journey planning behind each transport of laboratory animals.

A POP QUIZ

The workshop began with an interactive quiz moderated by Bruce Kennedy, compliance associate at California State Polytechnic University. Using an audience-response system called Poll Everywhere that would be used in future interactive sessions, even off-site attendees could answer

[1] Affiliations of speakers and participants were current as of September 2014.

Kennedy's questions using tablets or cell phones. Among the questions posed were:

1. It is freezing outside and you are ready to fly. The airline will transport your dog, too, in the hold of the plane. True or false?
2. Besides the obvious food and water, what else must we think about when transporting animals?
3. How many countries ship animals for research purposes?
   a. Fewer than 10?
   b. More than 75?
4. A group of nonhuman primates is to be shipped from overseas to an American university. Which agencies would have an interest?[2]

Kennedy and C. Ford Morishita, retired biology teacher and founding member of the National Academy of Sciences Teacher Advisory Council, noted that the workshop had been designed for attendees to provide questions, comments, and perspectives—feedback to be incorporated into the workshop organizers' final reflections.

OVERVIEW OF LABORATORY ANIMAL TRANSPORTATION

Drawing from decades of experience with Charles River Laboratories, where he served as corporate vice president for veterinary and professional services, William White provided a global overview of the animal transportation process. Laboratory animals are distinguished by their intended use in research, testing, or education, he said. While the term could encompass many different species, it is generally applied to animals with a

---

[2] *Answers:*

1. False; according to the Animal Welfare Regulations, animals shall not be subject to temperatures less than 45 degrees Fahrenheit for more than 45 minutes, unless accompanied by a Certificate of Acclimation from a veterinarian.
2. Temperature, oxygen, environmental safety, stress, bedding, and something to keep the animal warm.
3. More than 75. There are about 200 countries in the world, and about one-third of them ship research animals.
4. The U.S. Centers for Disease Control and Prevention; U.S. Customs and Border Protection; the Convention on International Trade in Endangered Species of Wild Fauna and Flora; the U.S. Fish & Wildlife Service.

defined health and genetic status that have been purpose-bred to be used in research, testing, or education. Laboratory animals rarely come from wild populations and are not intended to be used as pets.

The *Live Animals Regulations* (LAR), developed by the International Air Transport Association (IATA), list standards for about 7,000 species or container requirements.[3] However, less than 20 species compose more than 98 percent of the animals that are shipped or used in biomedical research. Mice, rats, and fish make up more than 85 percent of animals used in research worldwide.

Animals need to be transported from place to place for a number of reasons, said White. In some cases, they possess specialized anatomic, genetic, psychological, or metabolic conditions. Transportation enables researchers to share genetically unique strains and to collaborate on studies. Animal transportation eliminates the need to breed commonly used strains on site, which is logistically challenging and often difficult. Transporting cryopreserved germplasm—ova, embryos, and sperm—rather than live animals is another option, especially if complex, long-distance shipping is required. However, the receiving institution needs to be able to re-derive live animals, a time-consuming process with a variable success rate. Additionally, the institution needs to maintain the recovered animals at a desired health status. It should be noted that even if animals are reconstituted at a repository, they may still have to be transported to a final destination.

Safe transportation requires experience and thoughtful planning, White observed. Shipment is done either by combining air and ground transportation or entirely by ground. The overall goal is to minimize the risk of illness or infection, stress, and death or injury, but not every shipment is perfect. "There are miscommunications. Things will happen. . . . Traffic accidents happen every day," noted White. "The goal of this workshop is to [provide information that can help] minimize those occurrences." While regulations are designed to ensure that safe and reliable transportation is maintained, they may not necessarily produce optimal transport under all circumstances. To illustrate that animals' welfare is not necessarily every shipper's primary concern, White shared a series of photographs from the World Health Organization (WHO). The photos illustrated, among other things, a calf in the back seat of a car, multiple piglets in several crates on the back of a motorcycle, chickens grossly overloaded in a makeshift car-

---

[3] The guidelines are available at http://www.iata.org/publications/store/Pages/live-animals-regulation.aspx (accessed on September 14, 2017).

rier on the back of another motorcycle, and an unrestrained camel in the bed of an open truck. Another set of photographs illustrated deficiencies with containers, including an animal carrier with a door that had come off, a dog running around the cargo area, loose snakes, fish splashing about in water in the cargo space, and an escaped monkey.

The overall failure rate of transporting laboratory animals commercially is low, White observed. When comparing the number of containers with problems to all containers shipped by commercial breeders, only 0.07 percent of the laboratory animal containers fail. This number is likely inflated by several factors depending on the statistical calculations. "These [percentages] reflect more than a million containers in one year globally," White said. "While there [currently] is a reasonably good success rate, it's not perfect, but it will get worse if we start limiting the carriers, routes, and options."

White spoke of the challenges of ground transportation, including expenses, traffic, weather extremes, the risk of breakdowns, and many other issues that can result in delays and failures. Local and regional carriers are available, but there are not a lot of choices, and costs to use dedicated trucks are significant. Transportation requires the right equipment and knowledge and is subject to strict U.S. Department of Transportation (DOT) requirements and licenses. "The worst thing you can do is have a complex route over a long distance with many transfers," he said. "That is a recipe for failure."

White reviewed some of the terminology used in the transport of laboratory animals. The **consignor** or **shipper** is the institution that is shipping the animals. The **consignee** is the institution receiving the animals. The **carrier** is the company transporting the animals. In the case of air transport, the airline is the **primary carrier**. If a trucking company picks up or delivers animals to the airport, it is the **secondary** or **intermediate carrier.**

With regard to air transport, less than 40 percent of the global commercial air fleet is capable of carrying animals. Not all cargo compartments have appropriate environmental controls, and mixed loads typically contain various perishable cargoes. Many documents and approvals can be required for transport, and pilots or airlines can refuse to carry animals. The person or the organization shipping the animals is ultimately responsible for the microbiological status of animals transported by air, and the shipper can be responsible for escaped animals. Weather delays, temperature embargos, and cancelled flights all need to be anticipated, White said.

Journey planning, including anticipating how things can go wrong, is the only way to minimize the risks in shipping, White stated. Animal transportation is highly regulated, so shippers cannot make any assumptions about what is or is not required, especially when shipping internationally. Animals will experience some stress in transit, but there are too many variables to precisely control it, White said. When animals are shipped using a commercial carrier, the outside of the container and, rarely, even the animals could become microbiologically contaminated. Occasionally animals will become sick or die during or after transit, which may or may not be the result of errors in the shipping process. Complaining about problems does little to help, while working with the transportation provider to collect and analyze the facts can improve future shipments.

Laboratory animals and biological materials of animal origin will continue to need to be transported both within and between countries, said White. In the United States, many agencies and organizations regulate or otherwise influence this activity, including the U.S. Fish & Wildlife Service (FWS), U.S. Centers for Disease Control and Prevention (CDC), DOT, and U.S. Department of Homeland Security (DHS). In other countries, different agencies and rules apply, and sometimes requirements conflict. Species-specific requirements and prohibitions that are unique to individual countries may also exist. Differences may be as simple as the wording on health certificates or the outright prohibition of transport into or through a country of certain species or biological material.

In response to adverse transportation events, pressure from nongovernmental organizations, and a better understanding of animal needs and improved transportation practices, new regulations are continuously developed, which adds additional complexity to an already elaborate process, White noted. Increasing documentation, licensing, certification, and inspection requirements have added to the cost and time required for each shipment, which can be compounded by inconsistent interpretation of the requirements.

The bottom line, concluded White, is that transportation options for laboratory animals are diminishing. "There is a point at which some players chose to exit the scene," he said. "It is not personal. It is just business. It is cost. It is time. It is inconsistent interpretation of the requirements. It is increasing liability."

ORGANIZATION OF THE PROCEEDINGS OF A WORKSHOP

This Proceedings of a Workshop looks first at the transport of animals through the air cargo system (Chapter 2). It then examines the processes and requirements of land transportation (Chapter 3).

Factors influencing the health of animals during transportation and practical issues affecting the safety and well-being of animals in transit apply to all forms of transportation (Chapter 4). Individual species also have particular requirements, guidelines, and needs (Chapter 5).

Multiple regulators, guiding principles, and documents are involved in transporting laboratory animals (Chapter 6). Also, different stakeholders in the system have different perspectives on animal transportation (Chapter 7).

This proceedings closes with insights generated by two interactive sessions (Chapter 8) and with a compilation of key points from the workshop as well as a look to the future regarding transportation of laboratory animals (Chapter 9).

Appendix A provides the workshop agenda. Appendix B contains biographical sketches of the workshop organizing committee members and speakers. The various points made by the workshop speakers and discussed with the audience during the interactive exercises are compiled into a Transportation Checklist in Appendix C as a tool to help with journey planning of transporting laboratory animals. The workshop's statement of task is included as Appendix D.

Several chapters of the proceedings contain boxes that summarize recommendations made by speakers regarding the transport of laboratory animals. These should not be seen as conclusions of the workshop in general, by the members of the Roundtable on Science and Welfare in Laboratory Animal Use, the Institute for Laboratory Animal Research (ILAR), or the National Academies of Sciences, Engineering, and Medicine (the National Academies). They are included to further the workshop goal of providing useful information to the scientific community about matters relating to laboratory animal care and use.

Many of the speakers covered issues and procedures at a level of detail that cannot be reproduced in this workshop proceedings. Videos of the presentations and copies of the speakers' slides are available at the project website[4] and provide a wealth of additional information.

[4] See http://nas-sites.org/ilar-roundtable/roundtable-activities/transportation (accessed on September 14, 2017).

2

Air Transportation

Transporting animals by air provides specific capabilities and corresponding limitations. Gregg Pittelkow, who leads the global logistics team for research models at Covance Laboratories, Inc., and Carl Kole, an independent consultant and 40-year veteran of the aviation industry, offered an up-close view of the air transportation system. Bruce Clemmons, the manager of the FedEx Live Animal Desk, then examined how IATA sets standards for the carriage of animals. Finally, Robert Quest, the enforcement officer at the City of London Corporation, provided an international perspective on moving live animals around the world.

THE PROCESS AND REQUIREMENTS OF AIR TRANSPORTATION

Pittelkow began by pointing out that the worldwide commercial airline fleet exceeds 20,000 aircraft, a number that is expected to double in the next 20 years. The majority of this growth will be in small- to medium-sized aircraft and single-aisle aircraft.

The cargo capacity to ship research animals is limited, Pittelkow explained. Today, contrary to common belief, the cargo space in all jets is pressurized and heated to about 40 degrees Fahrenheit to prevent the freezing of luggage and cargo. Additional heating, ventilation, and air conditioning capacity vary by plane. When a company orders a new aircraft, it has choices about additional features, Kole explained. At a price tag of at least $100,000 per cargo compartment, however, airlines do not usually opt for additional air and ventilation unless they are in the animal transportation business. Pittelkow agreed, adding that in 2003, when he was responsible for the cargo compartment systems for Northwest Airlines, the ventilation and air conditioning on the Airbus A330 cost $285,000 and that price has gone up significantly since then. Not only are these initial costs high, but these systems weigh between 500 and 1,000 pounds or more.

That weight is carried every day for the life of the plane, whether there are animals aboard or not, adding as much as $20,000 per year to fuel costs.

Ventilation and air conditioning are not—in all cases—critical to transporting animals. However, they allow the transportation of more animals in a fixed amount of space than otherwise would be possible. Because heat is generated only when the aircraft engines are powered on, animals are exposed to the ambient air temperatures during loading and unloading and when the plane is not running. The temperature could vary greatly and potentially be extreme, depending on the season and the location of the airplane's departure and arrival.

In addition, the cargo holding area is generally small and is often located a distance away from the passenger terminal. When moving animals from their holding area onto the ramp, the airlines are required to stay within a 45-minute time window. This requirement ensures that animals are not exposed to temperature extremes for too long. But if the distance is substantial, the airlines may use the 45-minute limit as a reason for not accepting animal shipments. For these noted and other complicating factors, airlines may consider transporting animals to be both challenging and not cost effective.

Kole focused on the training that airline personnel receive regarding the many regulations governing the transport of live animals. These regulations vary greatly both by country and by economic region. For that reason, "it is important, especially if you are a shipper, that you have somebody specifically dedicated to that job, who is qualified to understand what the shipping process is, and what the rules and regulations are." Each country has a different training protocol and requirements. In the United States, training is relatively general, while is it more specific in the European Union (EU). In the United States, training is also "scattered" among a variety of agencies, including FWS, DHS, and the U.S. Department of Agriculture (USDA), among others. These agencies do not always communicate well with each other, which can cause complications and unintended consequences, especially when a carrier outsources to a third-party handler.

Not all personnel speak English, which can complicate training. Airlines sometimes assign a bilingual individual to be responsible for others who do not speak English, but this can become complicated and expensive. According to Kole, "without exception, the cost of training and compliance with existing regulatory requirements far exceeds revenue streams from transporting animals."

THE INTERNATIONAL AIR TRANSPORT ASSOCIATION AND AIR CARGO

Bruce Clemmons is chair of the IATA Live Animals and Perishables Board (LAPB). Founded in 1945, IATA is the industry leader for air transport in the world, and all major airlines in the world belong to the organization. Its mission is to represent, lead, and serve the air transport industry, delivering standards and solutions to ensure successful air transport.

Two groups within the LAPB help to develop the live animal regulations to function as industry standards, said Clemmons. One group is an advisory panel that includes representatives of the World Association of Zoos and Aquariums, the Association of Zoos and Aquariums in the United States, the Pet Industry Joint Advisory Council, the World Organisation for Animal Health (OIE), and the Animal Transport Association. The other group is an animal care team comprising industry experts. The IATA publication LAR is updated annually and provides guidelines based on input from airlines, industry stakeholders, trade groups, national and international organizations, and the animal care team and advisory panel. For example, LAR contains guidelines on how to measure crates for transporting animals, noting minimum specifications. Every shipper should discuss with the airline whether the container can safely fit on the aircraft, preferably before the container is built, Clemmons noted.

According to Clemmons, shippers (consignors) should be mindful that different countries have specific regulations as to what they can and cannot import or how animals can enter the country. He further noted that the Convention on International Trade in Endangered Species of Wild Fauna and Flora (CITES) approved LAR for non-air transport of CITES species. "This is an illustration of how well regarded our *Live Animals Regulations* are," he said.

CROSSING INTERNATIONAL BORDERS: A EUROPEAN PERSPECTIVE

Robert Quest is charged with ensuring compliance on the import and transit of animals at the Border Inspection Post at Heathrow Airport, which he manages. Most animals imported into London come through Heathrow (see Box 2-1), and Quest has a staff of about 30 people who work around the clock. Only a small percentage of the animals that come through Heathrow is laboratory animals and, for his staff, these animals are easy to care for, as they are sealed in a box. His staff cannot take the animals out, put them in a kennel, feed them, water them, or do other care tasks. However, laboratory animals are "quite an emotive issue," said Quest. "We have had demonstrations outside our place because we accept them."

BOX 2-1
Number of Animals That Come Through Heathrow Airport in a Typical Year

- Cats and dogs - 16,400
- Birds - 1,200
- Reptiles - 250,000
- Day-old chicks - 250,000
- Fish - 26 million
- Rodents/lagomorphs - 4,500
- Horses - about 300 (mostly racehorses)
- Butterflies - more than 1 million
- Beneficial insects - tens of millions

SOURCE: https://www.cityoflondon.gov.uk/services/animal-health-welfare /heathrow-animal-reception-centre/Pages/default.aspx (accessed on September 14, 2017).

Managing the number of live animals that come through Heathrow makes it imperative to have regulations governing their health and well-being, Quest observed. Airports worldwide must have similar rules for the same reason, he added, even where animal traffic is less than at Heathrow.

Quest described an extensive and overarching EU legislative framework, under which an EU directive is implemented in each member state according to national regulation, whereas an EU regulation is immediately implemented by all member states as is.

Quest discussed several common problems that arise at Border Inspection Posts (see Box 2-2). For example, shippers need to check the times when Border Inspection Posts are open in all the countries animals will be traveling through. "In Heathrow, we are open 24/7—but the state veterinarians who work with us aren't," he said. "Your shipment can come to us in the middle of the night at 3 a.m. That is not a problem, because we have people there. It won't get cleared, though, because the vets don't start work until the morning."

Public holidays are also not the same across the EU. Quest suggested that researchers visit the European Commission website, which contains extensive lists of information, including contact names, phone numbers, and emails. He also emphasized the importance of a contingency plan. In the United Kingdom, diversions can happen because of fog or other problems, and it is important to be able to contact both the consigner and the consignee.

BOX 2-2
Common Problems Encountered at Border Inspection Posts

- Poorly constructed boxes or crates (e.g., prevent air circulation when stacked)
- Lost documents and other paperwork problems
- Failure to consider time differences when a shipment arrives
- Failure to consider holidays when a shipment arrives
- "Non-usual animals" require different stocking densities (e.g., obese rats)

SOURCE: Quest slides 28 and 31, and spoken presentation.

Such problems highlight a fundamental theme of the workshop, Quest observed: Focused planning is necessary well before animals are shipped to consider possible issues and contingencies. For example, shippers need to consider all the stops animals must make along the way to ensure compliance with all the rules and regulations of each Border Inspection Post.

3

Land Transportation

Ground transportation plays a significant role in moving animals from place to place. Even when animals are transported by air, dependable ground transportation chauffeurs the animals to the airport and then, later, to their final destination.

Kenneth Kobus, director of logistics at Charles River Laboratories, provided an overview of the available options. Robert Fernandez, vice president of operations and quality assurance for Direct Services, Inc., then offered a close-up view of the environmental conditions in trucks and how qualified shippers care about the health and safety of the live animals they ship. Together, the pair communicated that choosing the right land carrier for animals requires as much thought as moving them by air.

Kobus elaborated on the many kinds of transit companies that shippers[1] might choose and the range of available services. Dedicated contract carriers, for example, often perform additional services for their customers, such as operating transfer facilities as animals make their way across the country. Pricing is generally high, because costs are based on mileage and types of vehicles, but also because of the special nature of the shipment, the timing, and the fact that special handling and ventilation are required. Specific requirements by a shipper—for example, the shipment be the only one on the truck—must be clearly noted on the contract, Kobus said.

Fernandez focused his discussion on temperature- and time-sensitive commodity shipments, one of which would be live animals. His company's goals are clear: Animals arrive alive and healthy, maintain the same biosecurity status with which they started, and arrive on time. Direct Services expects to meet all regulatory requirements while keeping its personnel safe. Transporting laboratory animals is always stressful for the animals—therefore, the goal is always to minimize their stress as much as possible, Fernandez said. Like airlines, ground carriers must comply with numerous regulations from USDA and other federal agencies, state

[1] For definitions and terminology commonly used in the transportation process, see page 4.

departments of health, state fish and wildlife departments, and other regulatory bodies.[2] However, the transportation of rats and mice, which are the most numerous laboratory animals, is not regulated by the Animal Welfare Act, Fernandez observed. Nevertheless, he noted, his company treats all animals as if they were regulated in order to ensure proper and humane treatment.

Fernandez reviewed many of the container requirements related to cleanliness, ventilation, and safety issues; food and water availability; and so on. Most importantly, he said, all transport containers must be guaranteed to contain the animal for the duration of the trip. He explained that animals are monitored regularly for any sign of distress and also must be sheltered from rain and snow, direct sunlight, and cold weather. His company ensures that the air in the cargo area allows normal breathing of all the animals being moved. Air temperature is carefully monitored, aiming for a range not to fall below 45 degrees Fahrenheit or above 85 degrees Fahrenheit.

According to Fernandez, USDA[3] requires a disaster plan be in place. Questions to consider when designing a disaster plan include what kind of disaster is likely to occur? Who will do what in case of emergency? The shipper of the animals is responsible for ensuring that the entire route is planned ahead of time and is safe for the animals. The shipper is also responsible for all the documents, health certificates, and permits that may be required by any state through which the animals travel. Adequate planning is especially critical when animals are being moved a long distance, such as across the country. Are teams of drivers in place? Are all the documents in order?

Fernandez said that ground transportation has many advantages. For example, a ground transporter is likely to have better environmental control than an airline. His trucks have a refrigerated unit and tighter temperature range. Noise is controlled. Some offer disinfected cargo areas. He added that safety requirements are having the effect of shutting down unsafe carriers.

Importantly, drivers for Direct Services have cell phones and GPS

[2] Detailed information on federal agencies and federal and state regulations is provided in Appendix C: Transportation Checklist.

[3] CFR [Code of Federal Regulations]. 2013. Title 9, Part 2. Animal Welfare Regulations; Subpart C, Research Facilities. Washington, DC: Office of the Federal Register. Available at https://www.aphis.usda.gov/animal_welfare/downloads/Animal%20Care%20Blue%20Book%20-%202013%20-%20FINAL.pdf (accessed on September 14, 2017).

tracking in case of emergency, or if they get lost. They also have temperature alarm pagers for when drivers are away from their vehicle in a restaurant or rest area. Ground transportation, however, can be more costly than air, Fernandez noted, especially for small shipments.

4

The Health Status and Logistics of Laboratory Animals

The session examined the impact of transportation on the health status of the transported animals. It began with an overview of critical junctures with hazardous potential, and concluded with an analysis of issues relating to transport of animals with commonly encountered health issues (e.g., pregnancy, genetic modifications).

Kathleen Pritchett-Corning, senior clinical veterinarian at Harvard University, spoke to attendees about the health status of animals before, during, and after transit. She defined an animal's health status as the presence or absence of various microbiologic agents monitored.[1] Specifically, Pritchett-Corning noted the junctures where healthy animals might risk infection by microbiological agents while being transported. Such agents can not only affect the animals' health but render them unusable for research. They also can have significant and long-lasting effects on receiving institutions that need to clean up those agents.

Pritchett-Corning suggested that shippers follow the law regarding maintaining the health status of animals during transport. National laws governing animal import exist to protect agriculture and human health. Often exemptions from some testing for laboratory animals exist, but consignors need to be sure that their paperwork and statements meet the legal and regulatory requirements outlined in the exemptions.

She reminded attendees that the health status of the animal is generally the responsibility of the consignor and not the carrier. She further pointed out that the best way to avoid risks of infection is to ship and/or receive animals transported by dedicated carriers. "If your carrier has controlled the shipment from door to door, as most of the major laboratory animal vendors do, then the risk is greatly reduced. Alas, as we've heard earlier in the morning, due to cost, logistics, and other things, door-to-door

[1] Also found here: http://www.ne-bsa.org/2014%20Sept%20NEBSA%20biosecurity%20from%20a%20rodent's%20point%20of%20view.pdf (accessed on September 14, 2017).

deliveries are unlikely to be possible for research institutions and universities."

According to the Hazard Analysis and Critical Control Point (HACCP)[2] principles, the greatest risk is that animals will change health status during transport and infect animals at the receiving facility. (Box 4-1 identifies control points where contamination risks may exist.) Importantly, such a change may not be immediately obvious. "We need to make sure that, when we receive animals, the integrity of the crate is maintained, that the crate is disinfected properly, and that we enter the crate in a way that maintains the disinfection and the integrity of the crate," she said.

BOX 4-1
Critical Control Points Where Potential Contamination Hazards May Exist

- Packing
- Transport to shipping vehicle (cart, car, truck)
- Holding facilities at airport or other locations
- Mode of transportation (truck, air)
- Transport to ultimate destination (truck, car, cart as it is wheeled through facility)
- Unpacking

SOURCE: Pritchett-Corning slide 9

Pritchett-Corning shared an experience she and her colleagues had with recurring mouse parvovirus outbreaks in a specific pathogen free (SPF) immunology facility. For 3 years in a row, mice were stricken with parvovirus toward the end of the year. Eventually, she and her colleagues deduced that wild rodents came in from the outside in late fall and early winter to avoid the cold. They would venture into warm places like airline shipping holding facilities, where laboratory animal crates were palletized, leaving behind feces.

"To sum up, there is a risk of health status change during transport for laboratory rodents, because you can't always control what is next to

[2] According to the U.S. Food and Drug Administration (FDA), HACCP is a management system in which food safety is addressed through the analysis and control of biological, chemical, and physical hazards from raw material production, procurement, and handling, to manufacturing, distribution, and consumption of the finished product (https://www.fda.gov/food/guidanceregulation/haccp) (accessed on September 14, 2017).

your rodents," she said. "You can't control for all these risks . . . but you can take precautions on arrival to protect your facility."

Lynn Anderson focused on the transportation of animals with special health-related conditions (see Box 4-2). Immunocompromised animals must be shipped in a container capable of excluding microorganisms, she noted. However, other than their increased susceptibility to infection and possible lack of insulating hair coat, requirements during shipment are the same as for other animals.

Recovery time required before a surgically modified animal can be shipped varies with the complexity of surgery and the animals' response to surgery. Animals need sufficient postoperative recovery to withstand the stress of shipment, Anderson said, which may be only 1 or 2 days for minor procedures but longer for more extensive operations. Animals with complications from surgery or requiring daily therapy should not be shipped, and animals with devices surgically implanted that have exteriorized components should be transported in individual containers or compartments.

Laboratory animals are typically exposed to environmental changes during transport, Anderson observed. These can affect both their light cycle and their food consumption patterns. Animals will often lose up to 10 percent of their body weight because they do not drink much during the first 12 to 15 hours of transit. Water provision is prescribed by regulations, but animals still need to be hydrated when they arrive at a facility.

Anderson talked briefly about nonhuman primates, noting that they are not wild caught but are purpose-bred, highly social animals with complex behaviors. Their shipping container should be constructed so that animals cannot escape, and so that food (the same food they are accustomed to) and water during transit can be provided without opening the animal compartment. In the case of an escape, only trained personnel should attempt to capture them as these animals can cause injuries.

BOX 4-2
Considerations in the Transport of Animals Listed by Lynn Anderson

- Animals that are clinically ill or have severe physical disabilities are unsuitable for shipment.
- Animals with inherited abnormalities or birth defects to which they have adapted, resulting in no apparent functional debilitation that affects well-being, may be shipped provided they have a health certificate describing the condition.
- Animals infected with agents of human or agricultural importance cannot be shipped without federal permits.
- Shipping near-term pregnant animals should be avoided (time-mated rodents may be an exception but even so there are risks).
- Shipment prior to embryo implantation can result in failure to implant in some species. (It is often better to ship a mid-term pregnant animal and have her deliver at the receiving institution.)
- Genetically modified animals may be less tolerant of environmental changes and may require drugs or specific diets.
- For obese animals, the number of animals per container may need to be reduced in warm weather.
- Separate species during transit to avoid fights, prey/predator behaviors, and attempts to escape enclosures.
- Allow time for surgically modified animals to recover before shipping.
- Containers used to transport animals should allow for observation.

SOURCE: Anderson slides 4–18. Further elaboration on these guidelines can be found in the International Air Transport Association (IATA) *Live Animals Regulations*, Chapter 12. Available at http://www.iata.org/publications/store/pages/live-animals-regulation.aspx (accessed on September 14, 2017).

5

Species-Specific Considerations

A series of presenters at the workshop addressed container design features, legal requirements and guidelines for containers and shipments, health requirements, and in-transit requirements for several species of laboratory animals. They also reviewed species-specific needs during transit.

NONHUMAN PRIMATES

Joe Simmons, a laboratory animal veterinarian and independent consultant with Insight Diagnostics and Consulting, gave workshop attendees a sense of the complexities of transporting nonhuman primates (NHPs). He, too, pointed to the need for good planning. NHP transportation is a trying exercise in planning and logistics, he said. Every detail must be thoroughly thought through and planned out, including permitting, both ends of ground transportation, air transportation, and preparation for weather. A research primate costs between $3,000 and $10,000 per animal, so a shipment of 1,000 animals can represent a total value in the millions of dollars.

Managing the numerous documents required is an essential part of shipping NHPs. In the United States, it often takes 3 to 4 months to receive a CITES export permit from FWS. Many countries have a pre-export quarantine process that lasts from 4 to 6 weeks and must be fulfilled before animals can be exported. The CDC quarantine process can be lengthy as well.

Weather, and especially temperature, can vary greatly along the route from city of export, refueling stop(s), city of arrival, and the ground transportation route. Animals may be leaving from a country with a temperature of 80 degrees Fahrenheit, but intermediate stops for refueling may be below zero. Furthermore, even if it is 100 degrees Fahrenheit on the tarmac at the city of arrival, the crew will be gowned, gloved, and wearing Tyvek suits. While regulations are strict about how long animals can be exposed to weather extremes, such rules do not exist for people, requiring that contingency plans be in place.

Sturdy crates that meet IATA standards are crucial, Simmons observed. He also pointed out that it is possible that a crate meets IATA standards but not the USDA and the FWS standards (see Box 5-1). A 2- to 4-kilogram cynomolgus macaque, for instance, requires a crate that is big, heavy, and carefully constructed of strong wood to ensure that the animal cannot chew through it.

The inspections required when animals come into the United States are very thorough. Simmons described his own experience with two FWS inspectors "who took a flashlight and looked at every single monkey—all 1,000 to 1,500 monkeys—*looking at every single eyeball*, making sure that every animal was okay. And they were. We had an inspector from CDC that day, as well, to make sure that the quarantine shipment was handled appropriately."

For the period from October 1, 2012, to September 30, 2013, CDC reported that 19,678 NHPs were imported into the United States with 2 recorded deaths, or 0.01 percent, said Simmons. "We would like it to be perfect, but it's getting pretty close," he said. "A lot of care and thought goes into transporting NHPs."

BOX 5-1
Some of the Organizations Involved in Overseeing the Transport of Nonhuman Primates

- The Convention on the International Trade of Endangered Species of Wild Flora and Fauna (CITES) issuing authority
- U.S. Fish & Wildlife Service (FWS) and its foreign equivalents
- U.S. Department of Agriculture (USDA) and its foreign equivalents
- Customs services from the exporting and importing countries
- U.S. Centers for Disease Control and Prevention (CDC) and its foreign equivalents
- U.S. Department of Homeland Security and the Transportation Security Administration
- State veterinarians and regulators

SOURCE: Simmons slide 3. Additional information regarding regulatory organizations can be found in Appendix C: Transportation Checklist.

DOGS AND FERRETS

Marshall BioResources is a global supplier of purpose-bred beagles, mongrels, ferrets, and guinea pigs raised in the United States, the United Kingdom, and China. Vice president Andy Smith echoed many of the themes of the workshop in urging careful planning, particularly with regard to the documentation required by various countries, states, and agencies (see Box 5-2). Smith also discussed the pros and cons of air versus ground transportation. The advantage of air transport is that the animals are in transit for the shortest period of time, whereas ground transportation provides better environmental control, he said.

Transporting dogs and similar animals involves many complex regulations, but all are designed around the animals' safety and well-being. Smith urged taking care with documentation and crate selection, understanding the risks, and preparing accordingly. Working with organizations that understand best practices can avoid many potential pitfalls.

Public perception is beginning to affect the transportation of research animals in general, and dogs in particular, Smith said. The number of airlines that will accept research dogs has declined steadily and significantly partially due to public relations. The chief executive officers of some airlines that have accepted animal shipments for decades have received literally tens of thousands of form letters by email in a 24-hour period. To avoid tarnishing their companies' image, they may make a decision to discontinue shipments of research animals.

"We've had situations where the pilot would refuse to accept a shipment of research dogs, because it's their personal opinion that dogs shouldn't be used in biomedical research," Smith said. While protesters may be well meaning, often animals will be in transit for much longer if they cannot fly. Ultimately, Smith noted, if animals cannot be moved to the destination where they are needed, research may move to where the animals originate, which is likely to be Asia.

BOX 5-2
Some of the Documents Needed to Accompany Shipments of Dogs and Ferrets

Domestic
- Animal health certificate
- Weigh bill (for air shipments)
- Bill of lading (for ground shipments)
- USDA transfer form that records the disposition of animals from the registered breeding facility to the research facility receiving the animals

International
- Airway bill
- Shipper certification
- Invoice
- Route plan
- Import permit
- Rabies vaccination certificate (often)
- Health certificate issued by a veterinarian accredited and endorsed by USDA

SOURCE: Smith slide 7. Additional information about documents required to ship laboratory animals can be found in Appendix C: Transportation Checklist.

MICE, RATS, AND SMALL MAMMALS

Keynote speaker William White of Charles River Laboratories also spoke about the transport of mice, rats, and small mammals. A wide variety of considerations go into the design, dimensions, construction, ventilation, labeling, and use of shipping containers (see Box 5-3). Containers that comply with IATA construction standards are available commercially. White urged those who are using IATA's LAR as part of their guidance to use the most current version of the guidelines.

Air shipment is highly regulated by government agencies and other oversight; it is not designed for the shipper's convenience, White reminded workshop participants. For example, animals need to get to the airport 2 hours ahead of time but not more than 4 hours ahead of time. Shipping internationally involves a maze of requirements, and failure to meet them may halt animals' movement at any point during the journey.

White mentioned the "mouse passport," an idea developed by the National Centre for Replacement, Refinement, and Reduction of Animals in Research (NC3Rs). This is not a legal document but a detailed packing list with assembly instructions and an operating manual.

BOX 5-3
Considerations Listed by William White in Transporting Mice, Rats, and Small Mammals

- Use a species-appropriate container—LAR container requirement 84, if a filtered container, or LAR container requirement 81, if a non-filtered container.
- Container must be large enough for animals to move about freely and make normal postural adjustments. They should have adequate airspace over the highest part of their body to allow air movement and to prevent injury from contact with the top of the container.
- Fully loaded crates must be able to stack at least eight high without damage or crushing of the bottom container.
- Interior surface must be smooth, moisture resistant, and durable. It must be covered by small grid wire with finished wire edges, solid smooth plastic, plastic film, or other materials that resist gnawing. The wire must not allow animals to gain purchase with their teeth.
- All critical junctions must be fastened with durable fasteners.
- Containers must have viewing windows.
- Floors must be designed to form a channel for liquids.
- Keep the inside of the container dry by wicking away moisture (e.g., place absorbent pads below the wire mesh).
- Do not reuse shipping containers.
- Do not reuse filters.
- Disinfect bedding right from the mill before using, then use plenty.
- Give rodents food pellets or extruded dry rodent food—same as in their home.
- Affix gel water packages to container by hot-gluing them to the inside of the container. Remember that gel water is not a nutritionally complete diet and containers need to provide food.
- When rodents run out of food and water, they look for a light source and will chew their way out. Provide food and water for an additional 24 hours, at least, in case of delays.
- Maintain adequate cross-ventilation following regulatory guidance.

- Mice can manage cold stress, but are heat intolerant. Annually updated IATA density guidelines need to be checked.

SOURCE: White slides 3–39 and spoken presentation.

FISH

Founded by the National Institutes of Health (NIH), the Zebrafish International Resource Center at the University of Oregon has the mission of collecting and distributing the zebrafish model worldwide. The use of fish has risen rapidly in popularity in recent decades. According to the Center's David Lains, this is because fish have external transparent embryos, fast life cycles and development, easily studied and manipulated genetics, and a number of transgenic lines. They are used not only in biomedical research but in evolutionary biology and ecology research as well. They can be kept at high densities and shipped during various stages of their life cycle. Adult fish can be moved faster, said Lains, but transporting embryos provides better biosecurity. While embryos are more cost effective, they are also more sensitive to temperature fluctuations and delays. Cryopreserved sperm is the safest and densest way of moving collections around the planet, he said.

Many considerations go into shipping fish (see Box 5-4). For example, Lains pointed out that animals need to be kept warm during transit, but for every 10 degrees Centigrade change, fish metabolism halves or doubles depending on the temperature zone. Because of this, Singapore is a much more dangerous location to ship fish than Iceland, even in the winter.

The export of zebrafish falls under the jurisdiction of FWS; USDA gets involved only if the recipient countries request health certification, and customs and border control are also involved. Lains recommended working with the carrier to develop the necessary document packages.

An underlying theme of the workshop was that unpredictable things can happen. An example was the explosion of the volcano Eyjafjallajökull in Iceland while a shipment of fish was en route to Ireland. The fish were rerouted hundreds of miles south. After multiple attempts to recover them, the authorities of the country where they arrived responded that the fish were "gone and not recoverable." Lains suspected they were being sold to tropical fish wholesalers. He sent a general inquiry asking how its news agencies would like a story about genetically modified fish being sold to children as pets. "Within days, the fish were in Ireland," Lains said. "Sometimes you have to think out of the box."

BOX 5-4
Considerations Listed by David Lains for Packaging Zebrafish for Transit

- Keeping water in the bag is critical but is easier said than done.
- Lains uses three-thousandths of an inch thick polyethylene watertight bags. The inner bag is handmade in a square format to avoid corners that can trap fish and a standard straight-bottomed outer bag.
- It is critical that the pleats at the top of the bag are even, because large overlaps can trap and kill fish.
- The bags are sealed with an aluminum clip, a fastener similar to those found on chicken bags or sausage casing. This clip helps form very tight bags that do not deform during shipping.
- In theory, rubber bands can replace clips, but putting rubber bands on properly is tricky, and if not done correctly, it could trap and kill the fish.
- Equal water and gas pressure within the bag is critical so the bag does not deform and trap fish in the corners.
- Boxes need to be labeled following IATA guidelines.
- Boxes need to be kept at room temperature.
- Use corrugated outer boxes with styrofoam inner lining (available through supply houses).
- Breathing stripes on the heat packs provide oxygen to the fish.
- Use heat packs to maintain even temperature in the winter. Completely surround the heat packs with tape so that cold air does not infiltrate the box.
- Isolate animals from the heat pack with a single layer of bubble wrap, which can be taped to the lid of the box.
- To fill voids in shipping packages, use non-biodegradable packing peanuts or fill extra bags with oxygen.
- Use water-filled fish bags with no gas to increase the thermal mass and moderate the temperature.
- During transit, the water pH drops significantly. Upon arrival fish need to be transferred immediately to clean water of similar condition to that of their origin.

SOURCE: Lains slides 9–16 and spoken presentation.

6

International, National, and State Regulatory Requirements

The second day of the workshop began by focusing on the regulatory oversight of transporting laboratory animals, with representatives of several regulatory agencies reviewing the guiding principles involved in transport. Each agency or organization has a unique role in the process and oversees different components of the transport, but the number and scope of regulations can generate uncertainty about successfully meeting the many requirements.

WORLD ORGANISATION FOR ANIMAL HEALTH

OIE, which is based in Paris, came into being almost 100 years ago, when a shipment of diseased cattle left India for Brazil. The cattle passed through Belgium, infecting European farms with rinderpest virus that killed countless animals and caused widespread devastation.

OIE has three basic goals, said P. Gary Egrie, farm animal welfare coordinator with USDA. The first is to convey information among countries to enhance safe trade. "If country X has a disease," said Egrie, "inform us so we can take actions to prevent those diseases from moving in international trade." The second is to coordinate responses among countries so that those responses are more effective. The third is to harmonize standards among the 180 member countries of the organization, representing most of the world's countries. OIE is designated by the World Trade Organization as the scientific reference body for animal health.

OIE's guidelines originally were developed around issues that Egrie called "low-hanging fruit," including animal transportation, slaughter, and stray dog control, through four commissions within the organization.[1] New mandates of OIE include animal welfare, animal food production safety, and veterinary infrastructure. Each commission includes a rotating group of specialists who ultimately decide what standards are adopted. The Biological Standards Commission, for example, would set the standards for how testing for tuberculosis is done. The United States has a representative on both the Terrestrial Animal Health Standards Commission and the Biological Standards Commission.

OIE has developed guidelines on the transport of animals by air, sea, and land. It also sets forth standards for the use of animals in research and education, Egrie noted.

U.S. CUSTOMS AND BORDER PROTECTION

Romelito Lapitan discussed his work with the U.S. Customs and Border Protection (CBP), the agency that oversees U.S. borders and ports of entry. According to Lapitan, currently the director of the Ag/Bio-Terror Countermeasures Division within the Office of Field Operations, Agriculture Programs and Trade Liaison, CBP is first in line, along with the agricultural specialists, to screen and process all foreign agricultural and biological imports into the United States; whether live animals, animal products, and animal by-products or propagative plants, plant products, and plant by-products. CBP assists 49 other federal government agencies enforce compliance to their respective regulatory requirements; such as CDC, USDA's Animal and Plant Health Inspection Service (APHIS), FWS, the U.S. Department of Transportation, and FDA in preventing entry of exotic pests and diseases that are harmful to American agriculture, wildlife and natural resources, and public health.

CBP inspects and ensures that all foreign agricultural imports and their carriers are legitimate, clean, and free of contaminants and prohibited products that may be harboring animal and plant pests and diseases. CBP's role on live animal importation, however, is limited to the review of the shipment manifest for accuracy and that all accompanying documents (e.g., import permit, health certificates) required by the regulatory agencies are complete and valid. Non-compliant live animal imports are referred

[1] Terrestrial Animal Health Standards Commission; Scientific Commission for Animal Diseases; Biological Standards Commission; Aquatic Animals Health Standards Commission. Available at http://www.oie.int/international-standard-setting/specialists-commissions-groups (accessed on September 14, 2017).

to USDA-APHIS Veterinary Services, CDC, and FWS for adjudication. Lapitan said that CBP does not have the facility to hold live animals at the port but will safeguard and take care of the animals until the regulating agency arrives, inspects, and recommends further action to perform. Regulatory actions may lead to re-exportation of the shipment back to its country of origin, transfer of custody from CBP to the regulating agency for further investigations, or released to U.S. commerce once compliance is met.

CBP operates in 328 ports of entry across the country—including 15 pre-clearance stations located in various regions of the world. CBP currently employs approximately 2,415 agricultural specialists who are distributed in 175 ports of entry, where large volume of cargo, foreign mail, and packages enter, and high volume of traffic in international travel is observed. According to Lapitan, not every port is equipped to receive every animal. Hence, APHIS and FWS restrict importation of livestock and wildlife, respectively, to ports designated as quarantine stations unless accompanied by a written port exemption permit to enter in non-designated ports. CDC requires importation of NHPs into the United States only through ports with CDC quarantine stations. Researchers need to keep this in mind when planning the route for their animals, he said. Also, beddings and food materials provided to live animals during transit are not permitted into the United States and, thus, are seized and destroyed.

U.S. FISH & WILDLIFE SERVICE

"If you take one thing away from this program today," said Sharon Lynn, Senior Wildlife Inspector at FWS, "it is to call the inspector at the port you intend to use to talk about your shipment." Lynn addressed attendees in her role as one of three senior inspectors stationed at FWS headquarters in nearby Virginia. According to Lynn, agencies have oversight over different issues. CDC, for example, deals with diseases affecting public health. USDA focuses on foreign plant pests, animal diseases, or bioterrorism agents. FWS focuses primarily on conservation issues. Lynn reviewed various IATA regulations, discussed the documentation relating to CITES and various other permits, and referenced the electronic Code of Federal Regulations (e-CFR[2]).

[2] See http://www.ecfr.gov/cgi-bin/ECFR?page=browse (accessed on September 14, 2017).

The FWS website lists all the designated ports with FWS inspectors, which account for just 18 out of the 328 official ports of entry across the United States. Furthermore, if the animals are also regulated by another agency, such as CDC, the port of choice must be staffed by that agency as well. In other words, the choice of ports of entry for bringing animals into the United States is severely limited. For example, Portland is a designated port for FWS but not for CDC; thus, NHPs cannot enter through Portland because it does not have CDC quarantine facilities. "Our inspectors do not have the authority to process or clear a shipment based on just an application," she said. "You must have a valid import/export license *before* your shipment shows up. And when you are dealing with live animals, time is of the essence." Customs clearance is received after FWS has approved the shipment, she said, again urging individuals to call the inspector at the designated port of entry. The inspector "can walk you through the required steps. It often depends on the country of origin. Some countries do not allow export of their wildlife."

U.S. CENTERS FOR DISEASE CONTROL AND PREVENTION

Gale Galland, who works part time with CDC's Division of Global Migration and Quarantine, talked to workshop attendees about CDC's role and authority in regulating the importation of live animals. The team's primary mission is to prevent and control the transmission of communicable diseases of public health importance. "When you bring in 1,300 animals and you have one sick, it affects the entire cohort, which means extended quarantine for up to 10 weeks," she said. "You can end up with some catastrophic losses."

Galland used the import of dogs, which she said must be healthy when they arrive at the port of entry, as an example. If any dog is ill, "we can require that it be seen by a veterinarian at the importer's expense," she said. This regulation was put in place to address concerns with rabies. Puppies must be at least 3 months old, but rabies vaccines must be given at least 30 days in advance, which means a 3-month-old puppy still needs to wait 30 days to be imported because rabies vaccines are not fully effective for about 1 month. Exceptions to this requirement are only granted for extenuating circumstances. Animals destined to be used in research, where rabies vaccination might interfere with the scientific outcomes, may be granted an exception.

Galland talked at length about various regulations regarding NHPs. The animals must undergo three negative tuberculin skin tests before they can be released. If they are sick or die, they have to be tested for filoviruses. In 1989, a group of animals coming into the country through Reagan International Airport in Washington, DC, were infected with what was subsequently termed Ebola Reston filovirus. While six people tested positive for the virus, no one got sick. "Everybody felt like we had really dodged a bullet here, so we began looking at our importers very systematically and inspecting animals at the airports."

As a result of the Ebola Reston outbreak, regulations involving shipments of NHPs were strengthened, including protective equipment standards for humans who are in close proximity to NHP crates. However, the situation is still not ideal, Galland said, showing recent photos with crates carrying NHPs nestled alongside other packages.

Galland highlighted other challenges, such as the limited number of airlines that will transport NHPs into the United States. Currently, no domestic airline will carry these animals; therefore, they must be transported by land to their final destination, which sometimes means crossing the country. Charter flights often incur scheduling delays and flight interruptions that place more stress on animals and increase the potential for animal illness and death during a long quarantine.

U.S. DEPARTMENT OF AGRICULTURE

Carol Clarke, research specialist with USDA's APHIS, spoke to the group about the role of APHIS-Veterinary Services. Veterinary Services deals specifically with animal health, through the enforcement of the Animal Health Protection Act. As a result, it only becomes involved with laboratory animals when agriculture may be impacted, such as the presence of a foreign disease like foot-and-mouth disease, or a finding of a select agent infection such as a highly pathogenic strain of avian influenza virus. In addition, fruits and vegetables in the animal enclosures are always a source of concern, as is bedding, especially wood chips from certain countries known to carry pests. APHIS-Plant Protection and Quarantine protects the nation's agriculture by ensuring the foodstuffs and bedding entering the United States do not contain harmful agents, said Clarke.

Carriers and intermediate handlers have specific requirements for transporting animals, Clarke said (see Box 6-1), including registering with USDA. Interestingly, there is no requirement for the consignor to provide a statement that either a dog or an NHP enclosure meets these requirements. Instead, the airline bears the responsibility for judging the adequacy of the enclosures for these two species.

Clarke referenced an amendment of the Animal Welfare Act affecting the import of dogs that went into effect in 2014.[3] With limited exceptions, this new rule prohibits the importation of dogs into the United States for resale, research, or veterinary treatment unless the dogs are in good health, have received all the necessary vaccinations, and are at least 6 months old.

BOX 6-1
Carrier and Handler Requirements for Transporting Animals from Carol Clarke

- A carrier cannot accept an animal earlier than 4 hours before the departure time. If needed, an extension for up to 2 hours may be granted.
- Carriers must provide verification that food and water were given to the animals within 4 hours before departure.
- Carriers are not permitted to transport an animal unless the primary enclosure meets the applicable requirements.
- Carriers and handlers must visually inspect animals at least once every 4 hours. Adhering to the 4-hour rule is challenging when animals travel by airplane; therefore, inspection should occur at least when loading and unloading.
- The ambient temperature for transporting animals should range between 45 and 85 degrees Fahrenheit. If exposure outside of that range is anticipated, a Certificate of Acclimation must be provided. If such a certificate is required, it must show the upper and lower temperature limits the animal can be exposed to and for how long. A veterinarian must sign the certificate.
- Accompanying paperwork must include the name and address of the consignor. There also must be verification that, indeed, food and water

[3] U.S. Department of Agriculture. Animal Welfare Act and Animal Welfare Regulations, 2013. Available at http://www.aphis.usda.gov/animal_welfare/downloads/Animal%20Care%20Blue%20Book%20-%202013%20-%20FINAL.pdf (accessed on September 14, 2017).

was given 4 hours before transportation and a statement that the enclosure meets the standards. If animals require any medications, that must be clearly stated. A number of other veterinary statements may be required as well.

- The crate or container with the animal must be clearly labeled "live animal" and be strong enough to pass the rigors of transportation and prevent the animal's escape.

- The crate or container must be designed to allow the animal adequate freedom of movement so the animal can stand up, turn around, and sit down.
- The crate should be designed to allow humans to quickly remove the animal in case of emergency.

SOURCE: Clarke slides 25–34 and spoken presentation. Additional detailed information on federal agencies, and federal and state regulations is provided in Appendix C: Transportation Checklist.

CONSIDERATIONS AT THE STATE LEVEL

The federal government regulates the entry of animals into the United States from other countries, as well as the interstate movement of animals, but states typically regulate the entry of animals into their own state, explained Daniel Kovich, program manager with the Office of Animal Care and Emergency Response, Division of Animal and Food Industry Services, Virginia Department of Agriculture and Consumer Services. Most states generally provide regulatory exemptions if animals are simply driven through the state, but in Virginia animals should be accompanied by a Certificate of Veterinary Inspection issued by a licensed and accredited veterinarian. Most states, however, do have a prohibited species list that includes dangerous, venomous, or invasive species. Large cats, for example, are on Virginia's prohibited species list. In such cases, importation may be allowed, but only for bona fide scientific purposes, and advance registration or permits are required.

Kovich reviewed the various terms, in addition to "agricultural animals," that Virginia uses in categorizing animals:

- "Game species" refer to native or naturalized wildlife.
- Research animals are considered as such only if they are un-

der the custodianship of a research facility and managed under a protocol approved by the facility's Institutional Animal Care and Use Committee (IACUC).
- "Companion animals" is a broad definition that includes—in addition to pets—lions, tigers, and circus animals such as elephants.

Most state laws and regulations respect and protect the use of animals in scientific research, but, as is the case at the federal level these protections are under ever increasing scrutiny. Animals expressly bred or sold for research purposes by USDA licensed breeders or dealers are considered companion or agricultural animals until the point of delivery to a research facility.

Kovich noted that Virginia's laws covering companion animal care are often viewed as a national model. These rules include provision of adequate food, water, and exercise; shelter that is properly cleaned; adequate space in the primary enclosure; and provision of care, treatment, transportation, and veterinary attention when needed. Many of the Virginia regulations mirror already existing federal regulations under the Animal Welfare Act.

While Virginia has many regulations governing animals imported into the state, Kovich explained that a particular requirement may be waived if directly tied to the aim of the research. That waiver, however, is given only to the research facility, so that facility must request the exemption prior to the animal being transported.

7

Perspectives on the Transportation of Laboratory Animals

The afternoon of the workshop's second day featured several perspectives on transporting laboratory animals, including those of an educator, a representative of a European communications and advocacy organization, an academic shipper, and a carrier.

INFORMING THE PUBLIC: AN EDUCATOR'S PERSPECTIVE

Retired science teacher C. Ford Morishita focused on the challenge of organizing an overwhelming amount of information about the use of animals in research and making it conveniently accessible to an audience that includes students, teachers, and the lay public. While websites of research institutions provide this information, tremendous patience is needed to wade through everything, he said. A better resource for the lay public might be websites of other organizations, such as ILAR, the Association for Assessment and Accreditation of Laboratory Animal Care (AAALAC) International, and USDA.

Morishita made the argument that the ultimate key to creating an informed public is through improving K–12 education. "How are we going to prepare those students by the time they leave high school?" he asked. Use of the Next Generation Science Standards (NGSS) is a particularly promising approach, he said, in that the standards promote a format of learning based on active demonstration of knowledge instead of memorization and regurgitation of content. NGSS builds a bridge among high school classrooms, higher education, and a research setting.

Morishita directed workshop attendees to the National Research Council's *Science, Medicine, and Animals*[1] and its accompanying teachers' guide. In particular, he provided an example from the teachers' guide that had been specifically crafted to discuss the Three Rs—the concepts of replacement, reduction, and refinement in the use of laboratory animals. During the lesson, students were given a scenario about the use of rats for testing in a pharmaceutical company. Students were asked to talk about the Three Rs and choose the most appropriate one for this case. They were asked to describe their answer, their reasoning, and why they would make a particular decision if they were the researcher, a member of the company advisory board, or someone in another role.

A EUROPEAN PERSPECTIVE

Kirk Leech, executive director of the European Animal Research Association (EARA), provided attendees with a portrait of activities in which animal rights groups are engaged to disrupt animal research throughout Europe. EARA is a communications and advocacy organization launched in February 2014 that seeks to uphold the interests of biomedical research across Europe.

While there has been a dramatic decline in criminal methods used to stop animal research, activists have effectively turned their energies to lobbying and social media to persuade companies not to transport research animals. It is often the threat of a Facebook or email campaign to a company's reputation and market share that will allow activists to achieve their goals.

Air France has become a major target for the anti-animal research groups, said Leech. The scientific community's support of Air France is important, as the actions of animal activists have affected transport providers and led many other companies, not only airlines, to stop transporting laboratory animals. "There are no longer any commercial maritime means available to transport animals in and out of the UK," Leech said.

Representatives of EARA recently met with representatives of the transportation sector who believe that this crisis results from a breakdown in trust between the public and the research community. Until that trust is

[1] National Research Council. 2004. *Science, medicine, and animals.* Washington, DC: The National Academies Press. Available at https://www.nap.edu/catalog/10733 (accessed on September 14, 2017).

won back, said Leech, the transportation of animals for research will not be resumed. "Because researchers are often very reluctant to talk about their research in public, the vacuum is filled by the opposition giving them the moral authority on this issue," he said. The public hears continually about the alleged misuse of NHPs, he said, but hears very few positive accounts to counterbalance these arguments. Researchers should be describing how the animals are cared for, said Leech, and why the research is important to find cures for diseases like dementia and Parkinson's.

Leech informed workshop attendees about an ongoing activist petition urging the European Parliament to reopen the debate about ending animal research. This petition, called the Citizens Initiative, has gained 1 million signatures thus far.[2] Leech described several initiatives that EARA has undertaken to counter these negative messages, responding in part to the requests of the transportation sector to speak publicly about the benefits of research. "The impact of all this on the European life sciences is quite dire," Leech said. "We need to find more imaginative answers to our problem."

AN ACADEMIC SHIPPER'S PERSPECTIVE

Steven Leary, assistant vice chancellor for veterinary affairs at Washington University in St. Louis, Missouri, has had the luxury of working with a full-time shipping coordinator for imports and exports of research animals. As part of an academic perspective on laboratory animal transportation, Leary emphasized the value of having such an expert within one's organization.

The shipping coordinator at Washington University provided animal resource services for between 400 and 450 principal investigators in the past 12 months alone. He processed 325 exports and 225 imports, with shipments going to about 160 locations around the United States and to about 20 other countries. Other large academic institutions in the United States, despite having comparably heavy traffic, have not centralized this process, leaving the tasks to various laboratory managers spread through-

[2] The European Commission's response to the Initiative can be found at: https://ec.europa.eu/jrc/en/news/commission-replies-to-european-citizens-initiative-against-animal-testing and at: http://ec.europa.eu/environment/chemicals/lab_animals/pdf/vivisection/en.pdf (sites accessed on September 14, 2017).

out the institution. "One couldn't expect these folks to have the necessary tools or understanding of the process," which compromises quality, compounds problems, and adds to the paperwork, Leary pointed out. Yet, researchers already are spending large amounts of time on administrative tasks—an estimated 42 percent, according to one recent study.[3]

Leary provided attendees with a quick overview of the many steps required to move animals. As described by earlier presenters, many people have to sign off on required forms. Problems with timing can be compounded by unrealistic expectations. Poor planning can cause delays and many other types of problems.

Labeling can also cause problems, as can weather and unanticipated accidents. Washington University once lost a box of mice when workers in an airport hangar literally ran over it. Another time, 6 out of 12 mice were lost because a driver took a circuitous route in Florida and the truck was not held at the proper temperature. "Some of these things are out of our control," Leary said. "With a coordinator and proper attention, however, these things can be minimized."

A CARRIER'S PERSPECTIVE

Carl Kole, former chair of the IATA LAPB, addressed the group again from the perspective of a consultant in the aviation industry. He talked briefly about some of the responsibilities of a carrier:

- Ensure that containers are species specific and compliant. Some kennels have "USDA-approved" or "airline-approved" tags. But airlines do not approve anything, said Kole, and neither does USDA.
- Provide space in the cargo hold that will support life—for example, by ensuring adequate ventilation.
- Provide storage facilities at both the origin and the destination.
- Separate laboratory animals from other animal shipments.
- Protect animals from the elements.

---

[3] National Science Board. 2014. *Reducing investigators' administrative workload for federally funded research.* Available at http://www.nsf.gov/pubs/2014/nsb1418/nsb1418.pdf (accessed on September 14, 2017).

According to Kole, and via the IATA Shipper's Certification for Live Animals, "carriers will not be liable for any loss, damage, or expense arising from death due to natural causes, or death or injury of any animal caused by the conduct or acts of the live animal itself or of other animals."[4] Nor are carriers liable for death or injury to an animal caused by defective packaging, or "by the inability of the animal to withstand unavoidable changes in its physical environment inherent to air transport."[5] Carriers do run into a problem with animals turning on each other, he said, mostly with dogs. "I often say to use a broker. They are experts. You are going to have a much, much better opportunity to get your animals where they need to go."

One of the greatest problems the industry encounters is that all the paperwork and all the security checks, while important, slow down the process. In fact, they slow things down so much, said Kole, that if the airline employees cannot get the animals from the warehouse into the airplane with the proper temperatures, the airline will simply refuse the shipment. Kole reminded attendees that airlines no longer operate for the public's convenience and necessity. While once a public utility, deregulations in 1982 changed the nature of the industry.

"You heard me talk yesterday about the economics of transporting animals," said Kole. "I don't think we are ever going to get over that hurdle. It costs a lot. That is not going to change. I truly hope we can come together collaboratively somehow, even if it is getting a group of diverse people to sit down and talk about what we can do to make your job easier—especially for the researchers."

---

[4] IATA Shipper's Certification for Live Animals. Available at https://www.aacargo.com/downloads/IATA_Shippers_Certification.pdf (accessed on September 14, 2017).

[5] Idem.

8

Interactive Sessions

During two interactive sessions at the workshop, David Kurtz, head of the Quality Assurance Laboratory, Comparative Medicine Branch, National Institute of Environmental Health Sciences (NIEHS), led the attendees in exercises designed to engage the audience in thinking about appropriate journey planning to support animal welfare during transportation. In the process, Kurtz derived a list of tasks individuals need to complete before transporting animals, including the questions and issues listed in Box 8-1.

BOX 8-1
Questions and Issues to Consider in Transporting Laboratory Animals

- If shipping rodents, can frozen embryos or sperm be shipped instead of live animals?
- If shipping live animals, what is their health status?
- What mode of transportation should be used?
- What is the quickest and most direct route with a minimal number of transfers?
- Have the animals been designated to be last on/first off when they are delivered to the plane?
- Does the airline have an emergency veterinary plan?
- Consider using an animal carrier to move animals (if evaluated and approved beforehand).
- Consider the calendar. Do not ship on weekends or holidays in the country of import or export.
- Consider the weather. What is the forecast for every stop of the transport? If a hurricane threatens the East coast, divert the shipment or hold off for a week or two.
- Check that the correct original documents have been prepared and signed with blue ink.
- Prepare an original document for every international border the animals will cross.

- Place documents in a place easily accessible by customs agents and others at the border.
- Be sure crates meet IATA guidelines, as well as FWS and CDC guidelines.
- Check that each animal has its own identification.
- Include extra food and water in case of delay.
- Check the bedding. Will it be accepted at all borders? For example, rodents sent to China cannot use the standard wood bedding used in the United States; use paper-based bedding instead.
- Ensure that lettering on the labels is at least one-inch high, thus meeting animal welfare regulations.
- Include phone numbers for all contacts. Ensure that key contacts are available by cell phone 24/7.
- Ensure that individuals who are handling the animals are trained and aware of protection practices, whether wearing gloves and a mask for handling rodents or full protective gear for handling NHPs.
- Develop an emergency rescue network based on the USDA's Animal Welfare Act Contingency Regulation. The Animal Transport Association has strongly endorsed this kind of planning.

SOURCE: Kurtz presentation. Additional detailed information on journey planning is provided in Appendix C: Transportation Checklist.

After walking attendees through a hypothetical shipping scenario and pointing out how complex such a scenario can be, Kurtz gave attendees a quiz, asking how they might proceed under different situations. In a manner similar to the interactive session that led off the workshop, this session allowed participation of on- and off-site attendees by means of Poll Everywhere. A sampling of the questions is below:

1. A new investigator is moving to your institution in Alabama from Canada and would like to bring his colony of 10 cynomolgus macaques with him. What is the first document that you should obtain for this shipment?
   - An airway bill?
   - A CITES permit?
   - A packing list?
   - An animal health certificate?

2. Can these animals be shipped directly from Toronto into Alabama?
3. Following a routine 31-day quarantine, the animals were released by CDC and FWS and are now en route to Alabama, 2 hours away. Midway along the route, however, you get a call from the truck driver. The truck has overheated and cannot continue. What should the driver do first?
    - Call for a tow to a local garage?
    - Pull into a parking lot and go to lunch?
    - Pull over to the side of the highway, check the animals, and call you, the consignee?
    - Pull off the highway to a discrete location, check the animals, and call his dispatcher?[1]

Kurtz ran through a number of other scenarios, explaining that this exercise was meant to illustrate the need to anticipate the unexpected. "Things will happen," he said. "Even though we plan as best as we can, we need to be prepared to act if something new arises."

---

[1] *Answers:*

1. Get the CITES permit first, because it takes a minimum of 90 days to process.
2. No, these animals cannot go directly to Alabama. Unless special permission is obtained from CDC and FWS, the animals must be imported through a designated FWS port. Furthermore, because they must be quarantined, they need to go to a CDC-approved quarantine site. Luckily Atlanta has both, so they can be sent to Georgia, traveling the rest of the way by ground.
3. Most people picked the last choice. It is a good idea, especially when transporting NHPs, to stop in a quiet, discrete location.

9

Key Points from the Workshop and a Look to the Future

The process of transporting animals is complex and time consuming. In order to capture the many practical suggestions provided by speakers and attendees, the leaders of the workshop compiled a list presented in Box 9-1. A more elaborate presentation in the form of a Transportation Checklist is included in Appendix C.

BOX 9-1
Practical Suggestions Discussed by Workshop Participants

- Check animals before shipping them to ensure they are healthy.
- Develop a written journey plan and checklist.
- Check the route. Transporting animals by the most direct route with fewer transfers is in the animals' best interest.
- Check the weather forecast before shipping animals. What will the ambient temperature be when they arrive at their destination? With air transport, might animals be stopping in Alaska in the winter or Florida in the summer?
- Check the calendar. Be sure to work around weekends and holidays in all countries where animals travel.
- Check the time zones. What hour of the day or night will animals arrive at their destination? Are veterinarians and others who need to examine your animals and documents working at that hour, or will animals be delayed?
- Check crates. Are they species specific? Will they pass the "Samsonite" test and prevent escapes?
- Check animals' supply of food and water. Is it sufficient for an extra day or two if transport is delayed? Are instructions included for feeding and watering the animals if needed? Animals are most inclined to escape when they are hungry.

- Check all states through which animals will be passing and their ultimate destination. Each state has different sets of requirements because each state is unique, with different animal populations, priorities, pressures, and so on.
- Check the qualifications of the veterinarian who signs off on animals' health. Some states require that veterinarians be *both* accredited and licensed if health certificates are to be valid.
- Consider whether animals can legally be moved or sold. Is a Material Transfer Agreement needed prior to their transport?
- Call ahead to every port of entry where animals will be stopping so that someone can specify what needs to be done in those locations.
- Call the state veterinarian's office in each state where business is conducted or animals are offloaded. The staff can answer questions or provide referrals if needed.
- Ensure that the individuals handling animals have the personal protective gear they need.
- Appoint a single point of contact at an institution. This person needs to be well trained, know the requirements and logistics, and ideally be the one who drafted the transportation plan. Train a backup in case the primary contact is unavailable.
- Do not underestimate the challenge of transporting animals, and work with experienced people who know what they are doing. Consider working with an animal carrier or broker. Some companies will handle every aspect of transportation. Several experienced companies are available, but they should be evaluated first. Ensure that they understand the needs of the species being shipped.
- Provide an original and a certified copy of all documents required for the transport. Be certain to include contact information for both the consignor (shipper) and the consignee (destination) in case a problem arises.
- Develop a contingency plan for every aspect of a journey, including the potential need to act quickly. Sometimes quick action is necessary.
- Be sensitive to the challenge to biomedical research posed by animal rights' groups and their influence on airlines.
- Be meticulous in determining the documentation needed to move animals. Start 4 months ahead of time if shipping internationally and if CITES documentation is required. Plan, plan, and plan.

CLOSING COMMENTS

In his closing remarks, Robert Dysko, professor and director of the Unit for Laboratory Animal Medicine at the University of Michigan Medical School, began by noting how valuable the workshop had been: "We are starting by educating ourselves, and then, hopefully, we will be able to go outside and start educating others."

Dysko, who also serves on the Board of Directors of the American Association of Veterinary Medical Colleges, believes the scientific community needs to continue to generate support for animal-based research and for the transportation needed to do that research. In part, this requires continuing a dialogue with airline companies in an effort to persuade more of them to transport laboratory animals. At the same time, a contingency plan needs to be established to replace commercial airlines as the primary means to transport animals across long distances, Dysko said. "Just pretending that some airline is going to turn around and start taking animals is not the way to go."

Dysko also talked about the "mind-boggling directives" of so many organizations with oversight of animal transport and the confusion in knowing what applies in any individual situation. No single answer fits every situation, he observed. Rather, every time the country, the species, or the route of transport changes, something different needs to be done. This complexity calls for the development of a handbook or educational materials that can guide all stakeholders in the transport of laboratory animals. In particular, such a handbook could serve as a guide to help shipping coordinators in academic and research institutions navigate the complex process of transporting animals.

In his closing comment, Dysko urged animal researchers to communicate with the public about their work, including the process of safely transporting the laboratory animals, which plays such a key role in identifying new therapies and curing diseases.

Appendix A

Workshop Agenda

Transportation of Laboratory Animals
A Workshop of the Roundtable on Science and Welfare in Laboratory Animal Use *(An Institute for Laboratory Animal Research [ILAR] Roundtable Series)*

September 3–4, 2014
2100 C Street NW, Washington, DC 20418
National Academy of Sciences Building, Room 125

Workshop Goals
This public workshop will examine critical issues relating to the transportation of laboratory animals. Invited speakers will address the challenges of transporting laboratory animals humanely and safely. Expected outcomes of this workshop include

1. Interactive sessions that aim to engage the audience in thinking about appropriate journey planning to support animal welfare while in transport.
2. Opportunities for the audience to discuss ways to improve transportation for laboratory animals.
3. A rapporteur-prepared summary of the presentations and discussions of the workshop.

Wednesday, September 3

7:30a.m. **Registration**

8:30 **Opening Remarks**
Lida Anestidou, Director, ILAR Roundtable
Lynn Anderson, Covance Laboratories, Inc., ILAR Roundtable Co-Chair

Morning Session - Moderator: Dianne Garnes, Novartis Corporation

8:45 **Transporting Live Animals: An Interactive Session**
Bruce Kennedy, California State Polytechnic University
C. Ford Morishita, Retired Biology Teacher/Founding Member of the National Academy of Sciences Teacher Advisory Council

This session will introduce the attendees to an audience response system that can be used with any tablet or cell phone device. Using this system (which will be used throughout the workshop), the audience will be queried about their opinions and knowledge regarding the intricacies of animal transportation. The session will also discuss how to use "the parking lot" concept.

9:05 **Overview of Laboratory Animal Transportation**
William White, Charles River Laboratories

This presentation will provide a global overview of animal transportation and examine current trends and process issues involving various transportation modes.

9:25 **How Animals Move Through the Air Cargo System**
Gregg Pittelkow, Covance Laboratories, Inc.
Carl Kole, Kole Consulting

This session will focus on the capabilities and limitations of transporting animals by air. The presenters will review relevant International Civil Aviation Authority requirements and operational practices, discuss necessary documentation and training, and analyze the need for safeguards to avoid cargo liability.

10:15 **Coffee Break**

10:30 **What Is IATA and How Is Air Cargo Controlled?**
Bruce Clemmons, FedEx Live Animal Desk

The International Air Transport Association (IATA) among its other responsibilities sets standards for the carriage of animals. This presentation will review the steps of this process including how standards are developed, implemented, and regularly reviewed; and how these standards impact other modes of carriage, and/or transportation guidance promulgated by national and international organizations.

10:50 **Crossing International Borders: A European Perspective**
Robert Quest, City of London Corporation

Moving animals internationally presents logistical as well as animal health and safety challenges. This presentation will examine the process of crossing borders in the European Union, address some of the difficulties encountered, look at oversight mechanisms, and discuss common errors in international transportation.

11:20 **Question & Answer Session - Speakers' Roundtable**
Bruce Clemmons, Carl Kole, Gregg Pittelkow, Robert Quest, William White

11:35 **Land Transportation: The Process and Requirements**
Kenneth Kobus, Charles River Laboratories
Robert Fernandez, Direct Services, Inc.

This session will provide an in-depth review of land transportation. It will focus on the equipment, requirements, documentation, availability, types of providers, and pertinent definitions relating to this mode of transit. It will also review the capabilities and limitations of the process (e.g., environmental control).

12:20p.m. **Lunch** (Will not be provided. A cafeteria is located on the lower level of the National Academy of Sciences building.)

Afternoon Session - Moderator: Judith Franco, Pfizer Inc.

1:20 **Health Status Before, During, and After Transit**
Kathleen Pritchett-Corning, Harvard University

This presentation will review the factors impacting the health of animals during the transportation process. It will discuss the different needs of animals from a research and a public health perspective (e.g., genetically modified animals) and outline mechanisms to prevent cross-contamination during transit, as well as at the point of receipt.

1:45 **Life Science Logistics for Laboratory Animals**
Lynn Anderson, Covance Laboratories, Inc.

This session will review practical issues impacting the safety and well-being of animals in transit: elements relating to behavior, physiology, clinical illness, and environmental conditions associated with the transport of laboratory animals.

2:10 **Question & Answer Session - Speakers' Roundtable**
Lynn Anderson, Robert Fernandez, Kenneth Kobus, Kathleen Pritchett-Corning

2:25 **Coffee Break**

2:40 **Species-Specific Presentations (20 minutes each)**
This series of presentations will address container design features; legal requirements and guidelines for containers and shipment; health requirements; and in-transit requirements for each of the species designated in the following sessions. Each session will also provide a review of species-specific needs during transit, including necessary care or biological/microbiological issues that need to be addressed.

Nonhuman Primates - Joe Simmons, Insight Diagnostics and Consulting, LLC

Dogs and Ferrets - Andy Smith, Marshall BioResources
Mice, Rats, and Small Mammals - William White
Fish - David Lains, University of Oregon

4:00 **Question & Answer Session - Speakers' Roundtable**
David Lains, Joe Simmons, Andy Smith, William White

Adjourn for the Day

Thursday, September 4

7:30a.m. **Registration**

8:30 **Welcome and Focus of the Day**
Carol Clarke, U.S. Department of Agriculture

Carol Clarke will provide a summary of Day 1 and introduce the themes of Day 2—namely, regulatory oversight, perspectives regarding the movement of laboratory animals, and interactive exercises regarding transportation planning.

Morning Session - Moderator: Judith Franco

8:45 **National and International Regulatory Requirements (15 minutes each)**
This series of presentations aims to familiarize the audience with the multiple regulators, guiding principles, and documents involved in transporting laboratory animals. Each agency or organization has a unique role in the process and oversees different components of the transport. While there is great complementarity, the number and scope of guidelines and regulations generate uncertainty about successfully meeting the many requirements.

World Organisation for Animal Health - P. Gary Egrie
U.S. Customs and Border Patrol - Romelito Lapitan
U.S. Fish & Wildlife Service - Sharon Lynn

U.S. Centers for Disease Control and Prevention - Gale Galland
U.S. Department of Agriculture - Carol Clarke

10:15 **Coffee Break**

10:30 **Considerations at the State Level**
Dan Kovich, Virginia Department of Agriculture and Consumer Services

Question & Answer Session - Speakers' Roundtable
Carol Clarke, P. Gary Egrie, Gale Galland, Dan Kovich, Romelito Lapitan, Sharon Lynn

11:00 **First Interactive Exercise: Journey Planning**
David Kurtz, National Institute of Environmental Health Sciences

12:00p.m. **Lunch** (Will not be provided. A cafeteria is located on the lower level of the National Academy of Sciences building.)

Interactive exercise will continue through lunch.

Afternoon Session - Moderator: Dianne Garnes

1:30 **Second Interactive Exercise: Journey Planning: "Houston, we have a problem."**
David Kurtz

2:30 **How to Inform the Public Regarding Animal Transportation: An Educator's Perspective**
C. Ford Morishita

During this session, participants will examine the need for effectively informing the public regarding transportation of research animals including opportunities to teach secondary and post-secondary students and facilitate a deeper understanding of this topic.

2:45 **Coffee Break**

3:00 **The European Perspective**
Kirk Leech, European Animal Research Association

This session will focus on a comparison of U.S. and European practices regarding animal transportation, an analysis of international collaborations meant to facilitate animal transport, and a discussion of common problems and challenges.

3:25 **Laboratory Animal Transportation: An Academic Shipper's Perspective**
Steven Leary, Washington University

The animal care and use program at Washington University (WU) in St. Louis, Missouri, serves approximately 400 principal investigators. The Division of Comparative Medicine coordinates approximately 225 imports and 325 exports annually to/from approximately 160 locations within the United States and 20 countries. This presentation will use the WU experience as a basis for discussion, including the role of the shipping coordinator, steps involved in the import/export processes including documentation types, information review, verification and approval, and troubleshooting common difficulties.

3:45 **It's Not Personal, It's Business: A Carrier's Perspective**
Carl Kole

The title of this presentation is meant to capture the obligations faced by a carrier, including but not limited to the logistics of animal transportation, the regulatory requirements that have to be addressed, and any common problems with commensurate solutions.

4:05 **Question & Answer Session - Speakers' Roundtable**
Carl Kole, David Kurtz, Steven Leary, Kirk Leech, C. Ford Morishita

4:20 **Meeting Summary and Steps Forward**
Robert Dysko, University of Michigan Medical School

Appendix B

Committee and Speaker Biographies

Lynn Anderson is the Vice President of Animal Welfare and Comparative Medicine for Covance Laboratories, Inc. A graduate of Iowa State University's College of Veterinary Medicine, she has more than 25 years of experience developing and directing animal care and use programs, having served as the Attending Veterinarian and, ultimately, as the Institutional Official for Merck Research Laboratories. She also provided leadership for Charles River Laboratories' global consulting and staffing business that provided technical and scientific personnel to academic, commercial and government research institutions. In addition to her clinical expertise, Dr. Anderson has extensive experience in animal facility design, personnel training, and management and regulatory affairs. She is a Diplomate and past President of the American College of Laboratory Animal Medicine and past President of the American Association for Laboratory Animal Science (AALAS) and the American Society of Laboratory Animal Practitioners. She is a former Chair of the American Board of Veterinary Specialties for the American Veterinary Medical Association. She also serves as a specialist consultant and Trustee for the Association for Assessment and Accreditation of Laboratory Animal Care, International, representing Americans for Medical Progress. In addition, Dr. Anderson served as the co-editor of *Laboratory Animal Medicine*, 2nd edition, and a member of the editorial board for the *Office of Laboratory Animal Welfare Institutional Animal Care and Use Committee Guidebook*, 2nd edition.

Carol Clarke received her bachelor's degree in the Natural Sciences from Johns Hopkins University and her DVM degree from the Tuskegee School of Veterinary Medicine. After receiving her DVM, she practiced small animal medicine in New York City for 13 years before entering the laboratory animal medicine training program at SmithKline Beecham Pharmaceuticals located in King of Prussia, Pennsylvania. Upon completion of the program, she entered NIH in 1998 as the Primate Facility Veterinarian for the Veterinary Resources Program. In 2001, she accepted a position with the Comparative Medicine Branch of the National Institute of Allergy and Infectious Diseases (NIAID) and became a Diplomate of the American College of Laboratory Animal Medicine in 2005. During her 10 years with NIAID, she served as Institutional Animal Care and Use Committees Coordinator, Vice Chair of the Rodent Gnotobiotic Committee, and Chief of Shared and Central Facility Operations. In addition, she prepared all USDA, Office of Laboratory Animal Welfare, and Association for Assessment and Accreditation of Laboratory Animal Care, International annual reports. Dr. Clarke accepted a position with USDA in 2011, and currently serves as the Research Specialist Staff Officer at APHIS-Animal Care Headquarters located in Riverdale, Maryland. Her duties include serving as the USDA Animal Care Representative for Public Responsibility in Medicine and Research, Interagency Coordinating Commitee on the Validation of Alternative Methods, and the National Veterinary Accreditation Program.

Bruce Clemmons is the Manager of the FedEx Live Animal Desk, which is responsible for approving and coordinating live animal shipments on FedEx scheduled services flights throughout the FedEx network. He has been a board member of the IATA Live Animals and Perishables Board since 1998 and has served as the Chair of the IATA Live Animals and Perishables Board since 2010. He is also on the Board of Directors of the Animal Transportation Association since 2005. Mr. Clemmons graduated from Illinois Wesleyan University in 1981 with a Bachelor of Arts degree in Political Science.

Robert C. Dysko has been a faculty member of the Unit for Laboratory Animal Medicine and the University of Michigan since 1990. During those two plus decades, he has had many major responsibilities for the Unit, including oversight of all campus animal facility design and construction projects, director of the rodent health surveillance program, membership on the university's and the Ann Arbor Veterans Affairs animal care and use committees, and director of the program for training graduate veterinarians in laboratory animal medicine and comparative medical research. In July 2012, he became the fourth Director in the 5-year history of the Unit for Laboratory Animal Medicine. He has been active in AALAS, serving on its Executive Board from 2008 to 2012, and in the role of President in 2011; the American College of Laboratory Animal Medicine, serving on its Board of Directors from 2000 to 2003; and the American Association of Veterinary Medical Colleges, beginning a 3-year term on its Board of Directors in 2013 as the at-large representative for the Department of Comparative Medicine.

P. Gary Egrie is from Long Island, New York, where he received his bachelor and master's degrees from Stony Brook University in 1990 and 1992, respectively. In the mid-1990s, he worked in Ecuador (his mother is Ecuadorian) raising larval shrimp for sale to shrimp farms. In 2004, he graduated from the University of Pennsylvania School of Veterinary Medicine. From 2004 to 2005 he worked at Michigan State University providing veterinary services for the Michigan Department of Natural Resources' fish hatchery program. In 2005, he took a position with USDA as a Veterinary Medical Officer in the Aquaculture Program, where he was involved with aquaculture-related programs, policies, and regulations. In 2009, he took a newly created position of Farm Animal Welfare Coordinator, but is still involved with aquatic animal health as it relates to animal welfare and OIE. Dr. Egrie used to have hobbies, but now he has three young children and just enjoys sleep when he can get it.

Robert Fernandez is the Vice President of Operations and Quality Assurance for Direct Services, Inc., a national logistics and transportation services company. He is a graduate of New York University, College of Business and Public Administration, where he received a BS in Accounting and Economics.

Judith B. Franco is Associate Director of Global Standardization and Business Resources for Comparative Medicine at Pfizer Inc. In this position, she orchestrates the laboratory animal supply chain and supply chain partners to maximize contract return and ensure access to lab models, supplies, and services for the Pfizer research community. Additionally, Dr. Franco manages the global supply chain for both nonhuman primate and canine resources, aligning supply and demand across research areas and lines within Pfizer. Dr. Franco also engages with external partners to promote the importance of biomedical research and ensure a positive climate for biomedical research. Dr. Franco received BS degrees in Biology and Environmental Sciences from the State University of New York at Plattsburgh in 1986. Since graduating, she has spent more than 27 years in the pharmaceutical industry, beginning with Ciba-Geigy at the Environmental Research Center in Farmington, Connecticut, where she held multiple roles in agrichemical toxicology testing. She joined Bristol Meyers Squibb in 1995 as an intern with the Veterinary Services group, where she developed her skills in laboratory animal medicine. Dr. Franco began her Pfizer career in 1996, and for more than 18 years she has delivered against the Comparative Medicine departmental goals while working in the Veterinary Science and Technology, Site Operations, and Administration and Business Resources groups. Dr. Franco is a long-standing member of AALAS and became a member of the Animal Transportation Association (ATA) in 2010. Dr. Franco currently serves as an ATA Board Member and as Chair of the ATA Laboratory Animal Transportation Committee.

Gale Galland graduated from the University of Georgia College of Veterinary Medicine in 1986 and after working for 2 years in private practice and research, she joined the U.S. Public Health Service (USPHS), working for the CDC. During her 21 years as a commissioned officer, she worked in a variety of positions, including Staff Veterinarian for the Division of Parasitic Diseases, Attending Veterinarian, and then Branch Chief for the Laboratory Animal Medicine Branch and lastly, for the Division of Global Migration and Quarantine (DGMQ) Zoonoses Team, which is responsible for preventing the importation of animals and animal products that pose a threat to human health. During her work with DGMQ, Dr. Galland utilized her expertise with nonhuman primates and worked in the Nonhuman Primate Import Quarantine Program, overseeing its importation for science, education, or exhibition. Later, she became the DGMQ Zoonoses Team Lead. In January 2013, Dr. Galland retired from the USPHS and currently works part time with DGMQ in the Nonhuman Primate Import Quarantine Program and part-time as a Clinical Veterinarian in private practice.

Dianne Garnes is a native New Yorker who attended Hunter College in Manhattan, where she obtained a BA in Biology. She then attended the New York State College of Veterinary Medicine at Cornell University and received a DVM degree. She joined a small veterinary practice in Hyattsville, Maryland, and worked there for almost 2 years. Then she left small animal practice to work as a Clinical Veterinarian in the laboratory animal facility at Georgetown University and was introduced to the world of research and academia. She is currently the Director of Animal Welfare

Compliance and the Animal Welfare Officer for Novartis Pharmaceuticals Corporation in East Hanover, New Jersey. In this capacity, she is responsible for the enforcement of all government regulations and Novartis policies that apply to animal welfare, and the oversight of the committee that establishes policies and reviews and approves the humane care and use of animals in research and development at the site. She has worked for Novartis (Ciba-Geigy before the merger) for more than 20 years with varied responsibilities, including Director, Laboratory Animal Services (1995–2000) and Director, Safety Pharmacology (2000–2005). In the latter position she had the opportunity to establish a telemetry unit, which performed Good Laboratory Practices (GLP) studies in validated animal models. Current and past professional affiliations include veterinary licensure in Washington, DC, and Maryland; Member of the American Veterinary Medical Association; and Board Member of the New Jersey Association for Biomedical Research.

Bruce W. Kennedy received his BS degree in Zoology and MS degree in Avian Sciences at the University of California, Davis. His career is a mix of animals, chemistry, and people that started in California and included 19 years in Virginia and Maryland, as well as travel in 30 countries on 5 continents. Most of it has been as a Research Scientist, conducting and managing studies with experimental animals in the disciplines of nutrition, physiology, and developmental biology. Mr. Kennedy started in lab animal science using coturnix quail for his graduate thesis in nutritional toxicology. He has also worked at the bench (analytical chemistry) with dogs in protein metabolism and rats in carbohydrate nutrition studies (USDA), writing GLP toxicology reports (Hazelton), and preparing experimental diets with test substances (FDA) and managing transgenic mouse colonies at NIH and the California Institute of Technology. Currently, he is a Compliance Associate at California State Polytechnic University, Pomona, administering both the lab animal and the human subjects research committees and assisting graduate students and PIs in their research efforts. Mr. Kennedy has been a teacher and trainer for many years, lecturing on lab animals science and education research at Cal Poly Pomona and lab animal management at AALAS Institute for Laboratory Animal Management. After receiving his Laboratory Animal Technologist certification, he was asked to inaugurate a lab animal training course for USDA. He obtained the Certified Manager of Animal Resources Certification in 2006 and the Certified Professional IACUC Administrator certification in 2009. He is currently enrolled in a doctoral program in educational leadership. Mr. Kennedy is past President of the Laboratory Animal Welfare Training Exchange and AALAS. He has served on the AALAS education and certification committees and sat on the Scientific Advisory Committee. He is a Director of the California Society for Biomedical Research. He is a recipient of the Bantin and Kingman Institute of Animal Technology Award, the AALAS George R. Collins Award for training and educating in laboratory animal science, and the Purina lab animal tech award. Currently, he serves as an ad hoc specialist with AAALAC, International.

Kenneth Kobus is the Director of Logistics for Charles River Laboratories. Mr. Kobus has more than 25 years of experience in Logistics Management across a broad range of industries, including life sciences. Prior to coming to Charles River, Mr. Kobus was the Director of Logistics for a global medical device company and previously, for a global biotechnology company. He earned a bachelor's degree in Logistics and Transportation Management and an MBA in Finance and Logistics from Northeastern University. He has obtained graduate certificates from Harvard University Extension School (Management), The Ohio State University (Logistics), and BioPharma Institute (Regulatory Affairs). Mr. Kobus is also a member of the faculty for University of Phoenix Online Campus, Graduate School of Business and Management, John G. Sperling School of Business. Mr. Kobus is a member of several professional organizations, including APICS, ATA, AALAS, American Society of Transportation and Logistics (AST&L), Council of Supply Chain Management Professionals (CSCMP), and Project Management Institute (PMI).

Carl B. Kole is a 40-year veteran of the aviation industry. His operational experience dates from 1968 to 1990. His work background as an airport agent to operations manager has provided him with a complete understanding of the operational issues facing the airport manager in today's environment. From 1990 to April 2008, Mr. Kole was the Administrator of Special Cargoes for United Airlines. In that role, Mr. Kole had the sole responsibility for determining, developing, and implementing processes and procedures for dangerous goods, pharma, live animal and perishable transport. Mr. Kole currently manages his own consulting firm (Kole Consulting) based in the Chicago area. Mr. Kole served as the Chairman of the IATA LAPB from 1994 to 2003, as Vice-Chair from 2003 to 2008, and participated as the Vice Chair and board member since 1990. Mr. Kole was also a member of the IATA LAPB and had been in that role from 1994 to 2008. As the Chair, Mr. Kole contributed and facilitated the writing of Chapter 17 of the *IATA Perishable Cargo Regulations*. The IATA regulations provide the worldwide aviation industry guidance and regulatory requirements on transport issues. His work with live animal transport and the harmonization of transport standards continues. This work was documented in the Universities Federation for Animal Welfare (UFAW) *Handbook on the Care and Management of Laboratory and Other Research Animals*. He continues to consult the IATA LAPB on an informal basis. Mr. Kole participates in various training and information presentation venues each year. Examples of previous trainings include those for the USDA Animal Care Inspectors in conjunction with the APHIS Preceptor program and the ILAR International Workshop "Meeting the Challenges of a Global Environment." Mr. Kole is recognized throughout the industry as an expert in his areas of expertise, which include shipping perishable cargo and cool chain management. Recent commendations by the Federal Aviation Authority (FAA), USDA, and AALAS attest to that expertise.

Daniel A. Kovich is the Program Manager of the Office of Animal Care and Health Policy at the Virginia Department of Agriculture and Consumer Services (VDACS). Dr. Kovich has primary responsibility for managing Virginia's various animal welfare programs, including animal pound and shelter inspections, the Animal Record Summary Database, the Dangerous Dog Registry, animal control officer training standards, and provision of veterinary technical services to local governments. Dr. Kovich also has responsibility for regulations promulgated by the Department pertaining to animal health and welfare. Prior to joining VDACS, Dr. Kovich served as a Supervisory Public Health Veterinarian for the USDA in Milwaukee, Wisconsin. He received his DVM and MPH degrees from the University of Minnesota, and a Bachelor of Science in Animal Science from Iowa State University.

David M. Kurtz received his veterinary medical degree from the University of Tennessee in 1989. After 2 years as a small animal private practitioner, Dr. Kurtz entered the residency program in Laboratory Animal Medicine at the University of Alabama at Birmingham (UAB) in 1991. Upon completion of his residency, Dr. Kurtz continued at UAB acquiring a PhD in Molecular and Cellular Pathology in 1998. Dr. Kurtz performed a postdoctoral fellowship at Washington University School of Medicine in St. Louis (WUSTL). His research focused on the regulation of metabolic gene expression by members of the nuclear hormone receptor superfamily of transcription factors. Dr. Kurtz also had an appointment in the Division of Comparative Medicine as a Clinical Laboratory Animal Veterinarian. He was promoted to research faculty in 2000 and was awarded research funding from the National Center for Research Resources (NCRR) under a Special Emphasis Research Career Award (SERCA–K01) and from the WUSTL Diabetes Research Training–Program Project. From 2003 to 2011, he served as the Attending Veterinarian at the U.S. Environmental Protection Agency National Health and Environmental Effects Research Laboratory in Research Triangle Park, North Carolina, under a contract with Experimental Pathology Laboratories, Inc. Dr. Kurtz received Diplomate status in the American College of Laboratory Animal Medicine (ACLAM) in 2005. During that same period, Dr. Kurtz also served as the Attending Veterinarian for The Hamner Institutes of Health Sciences and Integrated Laboratory Systems, Inc., both located in Research Triangle Park, North Carolina. Since 2011, Dr. Kurtz has served as a Staff Scientist in the Comparative Medicine Branch (CMB) of the NIEHS, 1 of 27 institutes within NIH. At NIEHS, his primary responsibilities include animal use protocol consultation and review, regulatory compliance, clinical laboratory animal medicine, and oversight of the animal health surveillance program. Beginning August 25, 2013, Dr. Kurtz assumed the role of the Head, Quality Assurance Laboratory within CMB at NIEHS.

David Lains has been the Aquaculturist and Sales Manager for the Zebrafish International Resource Center (ZIRC) at the University of Oregon since its inception in 2001. He obtained a bachelor's degree in Environmental Science/Zoology at The Evergreen State College, focusing on his life-long passion for aquatic environments and their inhabitants. He has an extensive background in aquatics and frequently consults on large and small-scale fish operations in both academic and commercial settings. In addition, Mr. Lains has a thorough knowledge of fish health and husbandry, and he routinely assists hobbyists, educators, and researchers worldwide with their questions and concerns. He builds and operates the aquatic life support systems at ZIRC as well as his own personal fish projects, which are primarily focused on heirloom Japanese goldfish.

Romelito Lapitan is a Program Manager at the Ag/Bio-Terror Countermeasures (ABTC) Division within the Agriculture Programs and Trade Liaison (APTL) Office, Office of Field Operations, CBP, DHS. He also served as Acting Branch Chief at ABTC and Ag/Bio subject-matter expert to DHS biothreat analysis and countermeasures, operations visualization, and CBP chemical, biological, radiological, and nuclear defense (CBRNE) programs. His current initiatives include developing tools, guidance, and methodologies for interdicting agricultural- and bio-terrorism resources, and handling and processing illicit trade of biologics at U.S. points of entry (POE). Before joining APTL in 2011, he served as an Agriculture Specialist and later, in a supervisory role, at the Otay Mesa Commercial POE in San Diego, California, where he enforced USDA regulations on all agricultural imports entering the United States from Mexico. He was also instrumental in improving the CBP application software ACE/M1 for processing U.S. trade imports in sea and rail environments. He holds a postgraduate degree in Environmental Biophysics and, prior to joining CBP in 2008, was affiliated with Colorado State University doing research with a focus on groundwater quality and atmospheric loading of greenhouse gases.

Steven L. Leary is the Assistant Vice Chancellor for Veterinary Affairs at Washington University in St. Louis, Missouri. Dr. Leary earned his DVM from Iowa State University and was a USPHS Postdoctoral Fellow in Laboratory Animal Medicine and Comparative Pathology at Johns Hopkins University. He is a past recipient of the American Veterinary Medical Association (AVMA) Charles River Prize, the AALAS Griffin Award, and the Iowa State University Stange Award. Dr. Leary has served as a member of the AAALAC Council, president of ACLAM, chair of the AVMA Animal Welfare Committee, AVMA Panels on Euthanasia and Human Slaughter, and the National Association for Biomedical Research Board. He lobbied for passage of the Animal Enterprise Protection Act of 1992 and has testified before the U.S. House Agriculture Subcommittee.

Kirk Leech is the Executive Director of EARA. EARA is a communications and advocacy organization seeking to uphold the interests of biomedical research across Europe. The creation of EARA was prompted by the need (expressed by the research community) to better inform the

European public on the continued need for, and benefit of, the human use of animals in biomedical research. Representing both public and private research organizations, the association facilitates collaboration among networks across the European scientific community in order to coordinate national efforts and provide accurate and reliable information to the public and decision makers regarding the importance of animal research. In doing so, EARA aims to improve understanding and encourage openness in animal research. Previously, Dr. Leech worked in government affairs for the Association of the British Pharmaceutical Industry. Prior to that, Dr. Leech worked for Understanding Animal Research (UAR), the United Kingdom's leading advocacy group on the use of animals in medical research. Before working with UAR, Dr. Leech acted as a consultant for the White House Writers Group, a strategic communications consultancy based in Washington, DC, and founded by a group of former U.S. Presidential speechwriters. Dr. Leech was engaged to advise clients on improving public opinion on the environmental, economic, and cultural impact of a new billion-dollar gold mine in Transylvania, Romania. Before this position, Dr. Leech advised Action Research in Community Health and Development, a tribal rights organization working in the eastern tribal areas on Gujarat, India, on influencing public opinion on the economic benefits of the Narmada Dam and in opposing the imposition of wildlife sanctuaries on tribal land. Dr. Leech is a regular writer and presenter to UK and European media with more than 200 articles and appearances on television and radio.

Sharon Lynn is a Senior Wildlife Inspector with the headquarters office of the FWS Office of Law Enforcement, where her work includes policy development and programmatic support to implementing laws that regulate the import and export of live wildlife and wildlife products from a conservation perspective. Before taking this position, she served as a Wildlife Inspector at the port of Chicago from 1992 through the end of 2007. Her work there included ensuring that live wildlife imports (including those being imported for research use) complied with conservation laws and humane transport requirements.

C. Ford Morishita is a retired science teacher of 33 years (2011), with 26 years served at Clackamas High School in Clackamas, Oregon. Assignments centered primarily on Biology, AP Biology, and Honors Biotechnology during his career. For the past 2 years, he has served as Science Specialist and Regional Science Coordinator at ESD112 in Vancouver, Washington. Mr. Morishita work focused on professional development design and delivery, with respect to state and national initiatives. This work included the NGSS, Common Core State Standards, assessment and evaluation, and overseeing the science materials center that supported a K–8 science cooperative, comprising 29 school districts in Southwest Washington. One area of responsibility was to address Office of Superintendent of Public Instruction regulations and practices related to laboratory animal dissection and proper handling and disposal of native and non-indigenous laboratory animals. This also included provision of alternatives to lab dissection practices in the classroom. Mr. Morishita has served on two consensus study committees for the National Research Council on Testing Teacher Candidates and Evaluation of the National Board for Professional Teaching Standards. He also completed a 5-year term as founding member of the Teacher Advisory Council for the National Research Council (NRC). During that time, Mr. Morishita worked in an advisory role on other NRC projects such as formal input and review of *Science, Medicine, and Animals: Teacher's Guide*, and *Enhancing Professional Development for Teachers: Potential Uses of Information Technology*. Moreover, Mr. Morishita served on the National Science Resource Center national advisory board from 2003 to 2009 (currently known as Smithsonian Science Education Center). In 2008, Mr. Morishita was one of only three classroom teachers, to be selected as a National Associate by the National Academy of Sciences for his service and contributions. Mr. Morishita received his MAT in Biological Sciences and BS in Biology from Lewis and Clark College. He was selected as the 1994 Presidential Award for Excellence in Science Teaching and 1997 Oregon Teacher of the Year.

Gregg Pittelkow received his bachelor's in Business Administration from the College of St. Thomas, St. Paul, Minnesota (1985). In 1982, Mr. Pittelkow began his long career in the airline industry, serving in a variety of positions in both Passenger Marketing and Cargo Operations at Republic Airlines, Northwest Airlines, and Delta Airlines. In 1994, he was given the opportunity to head up Northwest Airlines' fledgling passenger and cargo live animal programs. The changes he made to these programs greatly enhanced animal welfare, decreased citations and fines by more than 95 percent, and consecutively increased program revenues by more than 10 percent annually. In conjunction with his duties at Northwest, in 1994 Mr. Pittelkow was elected to the IATA LAPB, a position he served in until his retirement from Delta Airlines in 2010. While on the Board, Mr. Pittelkow initiated or oversaw a number of enhancements to the regulations, including approval to use standard plastic pet containers for species other than dogs and cats and creation of the airline industry's first standards for the acceptance and handling of time- and temperature-sensitive health care products. In recognition of his long service to the animal transportation industry, in 2009 Mr. Pittelkow was awarded the ATA International Award for outstanding contributions to the welfare of animals in international commerce. Since 2010, Mr. Pittelkow has remained active in animal transportation, serving as a consultant to airlines, government departments and agencies, and nongovernmental organizations. In 2013, he joined Covance Laboratories, Inc., where today he leads its global logistics team for research models.

Kathleen R. Pritchett-Corning received her BS and her DVM from Washington State University and completed her postdoctoral training in laboratory animal medicine at the University of Washington. She became a diplomate of the American College of Laboratory Animal Medicine in 2002. Dr. Pritchett-Corning was the Director of Research and Professional Services at Charles River Laboratories until 2013 and she is currently employed at Harvard University Faculty of Arts and Sciences as a Senior Clinical Veterinarian. She is also an Affiliate Assistant Professor in the Department of Comparative Medicine at the University of Washington.

Robert Quest obtained his degree at Cardiff University. He then did a spell teaching biology in Uganda before returning to the United Kingdom. For the past 29 years he has been an Enforcement Officer for the City of London Corporation, which involves ensuring compliance on the import and transit of animals at the Border Inspection Post at Heathrow Airport, which he manages. He is a member of the UK National Animal Health and Welfare Panel and chairs the regional branch, as well as sitting on various other relevant working groups. Mr. Quest is also employed as a government Wildlife Inspector (part time) for the Animal Health and Veterinary Laboratories Agency. His other specialty is the CITES regulations and identification of CITES species for UK Police and Customs. Mr. Quest has broad experience as a tutor, both in the United Kingdom and abroad.

Susan Brust Silk is the Director of the Division of Policy and Education in the NIH Office of Laboratory Animal Welfare (OLAW), where she oversees the interpretation of Public Health Service Policy on Humane Care and Use of Laboratory Animals regarding the use of animals in research, testing, and training at Public Health Service (PHS)-assured institutions. She develops and directs educational programs in the ethical and humane care and use of laboratory animals including the OLAW online webinar programs and the OLAW Web resources. Before joining OLAW, Ms. Silk worked at the NIH National Cancer Institute (NCI), Office of the Director, as the Senior Scientific Speechwriter and Special Communication Project Developer. She served the NCI Intramural Program as Senior Animal Policy Advisor and Director of the Office of Mice Advice. Ms. Silk has conducted research on murine plasmacytomagenesis at NIH NCI and the Karolinska Institute. She directed transgenic mouse core laboratories at both NIH and the Johns Hopkins University School of Medicine. Ms. Silk has an MS in Immunology/Genetics from the University of Maryland, a BFA in Design and Fine Art from the Maryland Institute, College of Art.

Joe Simmons pursued residency training in comparative medicine and a PhD in Veterinary Pathobiology, studying novel virus infections of laboratory animals at the University of Missouri-Columbia after completing veterinary school. He has served as a faculty member at the University of Missouri-Columbia, as a Research Veterinarian at a major pharmaceutical company, and as Director of Research Animal Diagnostic Services for Charles River Laboratories. In 2009, he joined Charles River Research Models Houston as General Manager, where he was responsible for import and supply of NHPs for Charles River Laboratories' internal and external customers. He is currently an Executive Consultant for Insight Diagnostics and Consulting. His primary areas of interest and responsibility include infectious diseases of NHPs and NHP transportation, biosecurity, and welfare.

Andy Smith is the Vice President at Marshall BioResources, a breeder of laboratory canines, ferrets, and minipigs with facilities in the United States, Europe, and Asia. He obtained his bachelor's degree in Biology from the State University of New York at Geneseo and went on to get a master's degree in Business Administration from the University of Rochester. Mr. Smith has been with Marshall for more than 20 years, working through various positions leading to his current overall responsibility for North American and Asian operations. As part of his role, he oversees all animal transportation-related activities. Mr. Smith is a long-standing member of the national AALAS organization and is the past President of the upstate New York branch.

William J. White received his VMD degree from the University of Pennsylvania (1970); his Master of Science degree in Laboratory Animal Medicine from The Pennsylvania State University (1972); and his Bachelor of Science degree from The Pennsylvania State University (1966). Prior to joining Charles River Laboratories, Dr. White was a tenured Associate Professor of Comparative Medicine of the College of Medicine at the Milton S. Hershey Medical Center, where he conducted basic research in a number of areas involving the effects of environmental variables on laboratory animals and laboratory animal anesthesia. In 1988, he joined Charles River as Director of Professional Services, subsequently holding a number of positions in the organization, and is currently Corporate Vice President for Veterinary and Professional Services. In this capacity, he oversees the corporation's worldwide diagnostic and professional services activities as well as its corporate biosecurity program. While at Charles River, he has continued to head corporate research programs in environmental factors influencing animal performance as well as other areas involving the care and use of animals in a research environment. He has authored or co-authored 75 peer-reviewed research articles or book chapters. Dr. White served on the ILAR committee that developed the 1996 *Laboratory Animal Management Guide for Rodents* and on the ILAR committee that developed the *Guide for the Care and Use of Laboratory Animals*. He co-edited the ACLAM text on anesthesia and analgesia in laboratory animals and has been a member of the editorial board of the journal Comparative Medicine. Dr. White is a Diplomate of ACLAM and the European College of Laboratory Animal Medicine (ECLAM). He is past President of ACLAM. He is a member of the International Association of Colleges of Laboratory Animal Medicine in which he holds the office of president. He is a member of IATA and serves on the Live Animals and Perishables Board as a member of its Animal Welfare team. He has played the lead role in the development of the new container standards for laboratory animals as well as in the development of the "Life Science Logistics for Laboratory Animals" chapter in the IATA LAR manual.

Appendix C

Transportation Checklist

CHECKLIST FOR THE TRANSPORTATION OF LIVE ANIMALS FOR RESEARCH

This checklist is provided to help you plan the transport of your animals for research. The following resources are for information purposes only and should not be considered complete. The checklist is not meant to replace official documents, permits, certificates, etc. It was created by compiling various points made by speakers and attendees of the Workshop on Transportation of Laboratory Animals.

SECTION A: General Information

Sending Investigator:	Receiving Investigator:
Sending Institution:	Receiving Institution:
Protocol Number:	Protocol Number:
Approval Date:	Approval Date:
Expiration Date:	Expiration Date:

Sending Institution Shipping Coordinator	Receiving Institution Shipping Coordinator
Name:	Name:
Office Phone:	Office Phone:
Mobile Phone:	Mobile Phone:
Email:	Email:

Backup Shipping Coordinator or Emergency After Hours Contact	Backup Shipping Coordinator or Emergency After Hours Contact
Name:	Name:
Office Phone:	Office Phone:
Mobile Phone:	Mobile Phone:
Email:	Email:

SECTION B: Species Information

Common name:

Genus species:

Strain or stock:

Total males:

Total females:

Number of animals with litters:

***Note: Transport of nursing animals is strongly discouraged.*

Age and number of animals in each litter:

- Provide purpose and veterinary authorization to transport these animals

Total number of offspring or larvae (list by age):

Total number of animals:

Health or physical conditions that may impact travel (e.g., age, obesity, immune status, surgical alteration):

- Provide purpose and veterinary authorization to transport these animals
- What provisions are being made to address these issues?

Number of shipping boxes:

Number of primary enclosures within shipping box *(for aquatic and semi-aquatic)*:

SECTION C: Initial Requirements

Do you have all of the following?

- ☐ A signed written agreement to transfer the animals between the two parties?
- ☐ A signed material transfer agreement where applicable?
- ☐ Funds to transport the animals?
- ☐ Written approval from the sending institution?
- ☐ Written approval from the receiving institution?
- ☐ The exclusion pathogen/agent/parasite list for the United States? (imports only)
- ☐ The exclusion/pathogen/agent/parasite list for the country of destination? (exports only)
- ☐ The exclusion pathogen/agent/parasite list for the receiving institution?
- ☐ Breeder's statement, if animal was born in the United States?

Have you consulted the following regulatory requirements or guidelines?

- ☐ **Regulatory requirements for transportation under the Animal Welfare Act (AWA)**
    - U.S. Department of Agriculture-Animal and Plant Health Inspection Service/Animal Care (USDA-APHIS/AC): General number: 301-851-3751
    - *AWA and Animal Welfare Regulations*: https://www.aphis.usda.gov/animal_welfare/downloads/AC_BlueBook_AWA_FINAL_2017_508comp.pdf
- ☐ **Requirements for Public Health Services (PHS)-funded studies**
    - National Institutes of Health-Office of Laboratory Animal Welfare (NIH-OLAW) 301-496-7163; https://grants.nih.gov/grants/olaw/olaw.htm

- *Guide for the Care and Use of Laboratory Animals*, 8th edition: https://grants.nih.gov/grants/olaw/Guide-for-the-Care-and-Use-of-Laboratory-Animals.pdf

☐ **Requirements for animals harboring select agents**

- USDA-APHIS-Veterinary Services/Ag-Select Agent Services: 301-851-3300 (option 3)
- CDC Division of Select Agents and Toxins: 404-718-2000; http://www.cdc.gov/phpr/dsat.htm
- Joint website for CDC and USDA: https://www.selectagents.gov

☐ **Transportation of animals with infectious agents**

- U.S. Department of Transportation (DOT) Pipeline and Hazardous Materials Safety Administration: 202-366-4433; https://www.phmsa.dot.gov

☐ **Transportation guidelines from the International Air Transport Association (IATA)**

- *Live Animals Regulations* (LAR): http://www.iata.org/publications/store/Pages/live-animals-regulation.aspx

☐ **Guidelines from the World Organisation for Animal Health (OIE)**

- *Terrestrial Animal Health Code*: http://www.oie.int/en/international-standard-setting/terrestrial-code

SECTION D: Type of Movement and Corresponding Requirements

IMPORTATION OF LIVE ANIMALS INTO THE UNITED STATES

Entry location:

Quarantine location (where applicable):

Federal agencies to be contacted depending on species:

- **U.S. Customs and Border Protection (CBP)**
  CBP clears all shipments entering the country:
  1-877-227-5511

Information Center:
http://www.cbp.gov/trade/basic-import-export
https://www.cbp.gov/travel/clearing-cbp/bringing-agricultural-products-united-states (provides specific clearing guidance for other government agencies)

- **U.S. Fish & Wildlife Service (FWS)**
  FWS has jurisdiction over most animals, whether wild or captive-bred. Animals considered domesticated under 50 CFR 14.4 are not regulated by FWS. If any animal listed in the regulations is crossed with CITES-protected animals (e.g., hybrids), the animal falls under FWS jurisdiction. Also, animals on this list that become CITES protected are no longer considered domesticated.

  Contact FWS at the port the animals are being imported.
  https://www.fws.gov/le/inspection-offices.html
  - *The port used must be a designated port for both USDA and FWS.*

  Wildlife Inspection Office at the port of entry:
  https://www.fws.gov/le/designated-ports.html

  Additional information:
  Commercial Import/Export License:
  https://www.fws.gov/le/le-permits.html

  Division of Management Authority (only if species is protected under CITES): 703-358-2093 or 703-358-2104;
  https://www.fws.gov/international/permits

- **U.S. Centers for Disease Control and Prevention (CDC)**
  National Center for Emerging and Zoonotic Infectious Diseases (NCEZID)
  Division of Global Migration and Quarantine (DGMQ):
  404-498-1600

CDC has specific import requirements for nonhuman primates, African rodents, bats, civets, and their products. In general, CDC has jurisdiction over any live laboratory mammal and its products (e.g., blood, DNA, enzymes) that was exposed to or carries an agent that can be a threat to human health. The import permit either declares the animals are free of, or have not been exposed to, agents that can pose a threat to human health or indicates they have been treated to render such agents non-infectious.

See import program (IPP) for details: https://www.cdc.gov/phpr/ipp/index.htm

Questions can be directed to the import team: cdcanimalimports@cdc.gov

General information: http://www.cdc.gov/animalimportation/bringinganimaltous.html; 1-800-232-4636 or TTY: 1-888-232-6348

- **U.S. Department of Agriculture-Animal and Plant Health Inspection Service**
  Veterinary Services - General Information: https://www.aphis.usda.gov/aphis/ourfocus/animalhealth/contact-us

  National Center for Import and Export: 301-851-3300; http://www.aphis.usda.gov/wps/portal/aphis/ourfocus/importexport

  Plant, Protection, and Quarantine (PPQ): 301-851-2046; http://www.aphis.usda.gov/wps/portal/aphis/ourfocus/planthealth (ensures foodstuffs and bedding provided for the animal during transport are free of harmful agents to U.S. agriculture)

  Animal Care: Importation of Live Dogs: 301-851-3751

- **National Oceanic Atmospheric Agency (NOAA)**
  NOAA Fisheries-National Marine Fisheries Service
  301-427-8400; http://www.fisheries.noaa.gov

EXPORTATION OF LIVE ANIMALS TO A FOREIGN COUNTRY

1. U.S. site of embarkation:

2. Destination country:
   - Obtain requirements through the consulate, embassy, or ministry of agriculture
   - Documents may need to be written in the destination country's native language

3. All borders crossed before reaching final destination:
   - Applies to land travel or off loading/hold over from a flight before continuing on
   - Information can be obtained from each country's consulate, embassy, or ministry of agriculture

Federal agencies to be contacted prior to exportation depending on the species

- **U.S. Customs and Border Protection (CBP)**
  CBP clears all shipments that exit the country;
  1-877-227-5511
  Information Center:
  https://www.cbp.gov/trade/basic-import-export
  https://www.cbp.gov/travel/clearing-cbp/bringing-agricultural-products-united-states (provides specific clearing guidance for other government agencies)

- **U.S. Department of Agriculture-Animal and Plant Health Inspection Service**
  Veterinary Services - General contact information:

https://www.aphis.usda.gov/aphis/ourfocus/animalhealth/contact-us
Veterinary Services National Center for Import and Export: 301-851-3300; https://www.aphis.usda.gov/wps/portal/aphis/ourfocus/importexport

- **U.S. Fish & Wildlife Service (FWS)**
  FWS has jurisdiction over most animals whether wild or captive bred. Animals considered domesticated under 50 CFR 14.4 are not regulated by FWS. If any animal listed in the regulations is crossed with CITES-protected animals (e.g., hybrids), the animal falls under FWS jurisdiction. Also, animals on this list that become CITES protected are no longer are considered domesticated.

  Contact FWS at the port the animals are being imported. https://www.fws.gov/le/inspection-offices.html
    - *The port used must be a designated port for both USDA and FWS*

  Wildlife Inspection Office at the port of entry: https://www.fws.gov/le/designated-ports.html

  Additional information:
  Commercial Import/Export License: https://www.fws.gov/le/le-permits.html

  Division of Management Authority (only if species is protected under CITES): 703-358-2093 or 703 358-2104; https://www.fws.gov/international/permits/www.aphis.usda.gov/wps/portal/aphis/ourfocus/importexport

- **National Oceanic Atmospheric Agency (NOAA)**
  NOAA Fisheries-National Marine Fisheries Service: 301-427-8400; http://www.fisheries.noaa.gov

INTER- AND INTRA-STATE TRANSPORTATION INVOLVING THE CONTINENTAL UNITED STATES, HAWAII, ALASKA, AND U.S. TERRITORIES

State and Territory Requirements

Health, vaccination, and testing requirements:

- Contact the State Animal Health Official (SAHO)
  U.S. Animal Health Association
  4221 Mitchell Ave.
  Saint Joseph, MO 64507
  816-671-1144
  http://www.usaha.org/federal-and-state-animal-health

Permits and species restrictions:

- Contact FWS
  https://www.fws.gov/offices/?ref=topbar

Prohibited/invasive species:

- National Agricultural Library-National Invasive Species Information Center
  https://www.invasivespeciesinfo.gov/resources/lists4states.shtml

State anti-cruelty laws/statutes:

- http://www.animallaw.com/laws.cfm

SECTION E: Journey Planning

1. **Dates of movement**

2. **Origin of the animals**

3. **Final destination of the animals**

4. **Are the animals being imported into the United States or exported from the United States?**
    - Borders crossed (ground transportation)

5. **For import or export, has the FWS port been contacted to arrange inspection?**

6. **Are all import or export documents finalized?**
   - Do you have one original copy for each border crossed?
   - Are documents required to be written in the destination country's native language?
   - Are documents attached to container/caging for easy inspection?

7. **Can the bedding be imported into the state or country of destination?**

8. **Is the enclosure properly labeled?**

9. **For airline travel:**
   - Are non-stop flights available?
   - Are the flights confirmed?
   - Is the pilot alerted to animals being included in the manifest?

10. **At what time will the movement begin?**

11. **At what time are the animals expected to arrive?**
    - Is there a time difference between the site of origin and the site of destination?

12. **What is the total travel time of the trip?**

13. **Will the travel occur over the weekend?**

14. **Will the travel occur during a U.S., state, or foreign country holiday?**
    **Note: This type of travel is highly discouraged.

15. **What are the weather reports along the entire route?**

16. Who are the points of contact for each leg of the trip?

17. Will an intermediate handler take possession during the course of travel?

- If so, when?
- Is there a written agreement?
- If multiple intermediate handlers are involved, list them:

18. Is the operator of the transport vehicle adequately trained to transport the species?

19. Will transportation be provided by a privately owned vehicle?

- If so, what is the person's name?
- Is the vehicle temperature- and humidity-controlled?
- Are backup heating, ventilation, and air conditioning (HVAC) systems available?
- How are environmental conditions monitored and recorded?
- Have the state regulations regarding transport of animals in private vehicles been reviewed (e.g., anti-cruelty always, prohibited species, commercial licensure agreements)?

20. How often will the animal be observed during the trip by the carrier or intermediate handler? (*Note: may not be applicable to aquatic species*)

- Is there a written agreement?
- How will the observation be recorded?

21. How often will the animals receive water during the trip? (*Note: not applicable to aquatic species*)

- How will water be provided (e.g., water, gel, other fluid, and foodstuffs)?

SECTION F: Planning in Case of Emergencies and Other Unexpected Events

1. **How is the shipment tracked during transport?**

2. **Can operators of the transport vehicle communicate the status of the shipment to senders and receivers during travel?**
    - What is the mode of communication (e.g., cell phone, CB radio, walkie-talkie)?
    - What are the operators' cell phone numbers?

3. **Is there enough food and water for 48 hours beyond the duration of the trip?**

4. **Is the recipient institution prepared to receive early or delayed shipments?**

5. **What back-up transportation is available in the event of a traffic accident or mechanical failure?**

6. **Where can the animals be housed in the event of a mechanical failure or stranding?**
    - In the vehicle, an institution, animal shelter, or zoo?
    - Is there a minimum/maximum temperature required for the holding area?

7. **What measures will be implemented to ensure animals are checked, fed, and/or watered during the delay?**

8. **What arrangements are there for sick or injured animals during transport?**

9. **Is there a rescue network (transport and veterinary) available along the transport route?**

10. **Is there a preparedness or response plan in the event there is an escape?**

Appendix D

Statement of Task

An ad hoc planning committee will plan and conduct a public workshop to examine critical issues relating to the transportation of research animals. Invited speakers will address the challenges of transporting laboratory animals humanely and safely to ensure adequate supply for biomedical research in the face of increasing societal opposition to the domestic and international movement of these animals. The ad hoc committee will develop the workshop agenda, select and invite speakers and discussants, and moderate the discussions. An individually authored summary of the presentations and discussions at the workshop will be prepared by a designated rapporteur in accordance with institutional guidelines.